NEW DIRECTIONS IN RURAL TOURISM

New Directions in Tourism Analysis

Series Editors: Kevin Meethan, University of Plymouth
                    Dimitri Ioannides, Southwest Missouri State University

Although tourism is becoming increasingly popular as both a taught subject and an area for empirical investigation, the theoretical underpinnings of many approaches have tended to be eclectic and somewhat underdeveloped. However, recent developments indicate that the field of tourism studies is beginning to develop in a more theoretically informed manner, but this has not yet been matched by current publications.

The aim of this series is to fill this gap with high quality monographs or edited collections that seek to develop tourism analysis at both theoretical and substantive levels using approaches which are broadly derived from allied social science disciplines such as Sociology, Social Anthropology, Human and Social Geography, and Cultural Studies. As tourism studies covers a wide range of activities and sub fields, certain areas such as Hospitality Management and Business, which are already well provided for, would be excluded. The series will therefore fill a gap in the current overall pattern of publication.

Suggested themes to be covered by the series, either singly or in combination, include – consumption; cultural change; development; gender; globalisation; political economy; social theory; sustainability.

Also in the series

Tasting Tourism: Travelling for Food and Drink
Priscilla Boniface
ISBN 0 7546 3514 7

Tourism and Economic Development
Case Studies from the Indian Ocean Region
Edited by R.N. Ghosh, M.A.B. Siddique and R. Gabbay
ISBN 0 7546 3053 6

Tourist's Experience of Place
Jaakko Suvantola
ISBN 0 7546 1830 7

New Directions in Rural Tourism

Edited by
DEREK HALL, LESLEY ROBERTS and MORAG MITCHELL

*with all good
wishes and
thanks
from
Derek Hall
21/1/04*

ASHGATE

Published by
Ashgate Publishing Limited
Gower House
Croft Road
Aldershot
Hants GU11 3HR
England

Ashgate Publishing Company
Suite 420
101 Cherry Street
Burlington, VT 05401-4405
USA

Ashgate website: http://www.ashgate.com

British Library Cataloguing in Publication Data
New directions in rural tourism. - (New directions in
    tourism analysis)
    1. Tourism 2. Rural development
    I. Hall, Derek, 1948- II. Roberts, Lesley, 1954-
    III. Mitchell, Morag
    338.4'791'091734

Library of Congress Cataloging-in-Publication Data
New directions in rural tourism / edited by Derek Hall, Lesley Roberts and Morag Mitchell.
        p. cm. -- (New directions in tourism analysis)
    Includes bibliographical references and index.
    ISBN 0-7546-3633-X
    1. Tourism. 2. Rural development. I. Hall, Derek R. II. Roberts, Lesley, 1954- III.
    Mitchell, Morag. IV. Series.

    G155.A1N428 2003
    338.4'791'091734--dc22

                                                            2003054487

ISBN 0 7546 3633 X

Printed and bound by Athenaeum Press, Ltd.,
Gateshead, Tyne & Wear.

Contents

List of Figures   *vii*
List of Tables   *viii*
List of Contributors   *ix*
Preface   *xi*
List of Abbreviations   *xv*

PART 1: CONTEXT

1   Tourism and the Countryside: Dynamic Relationships
    Derek Hall, Morag Mitchell and Lesley Roberts   3

PART 2: CONCEPTUALISATION

2   New Directions in Rural Tourism Impact Research
    Steven Boyne   19

3   Rural Tourism and Sustainability – A Critique
    Richard Sharpley   38

4   What is Managed when Managing Rural Tourism? The Case of Denmark
    Anders Sørensen and Per-Åke Nilsson   54

PART 3: EXPERIENCE

5   Encouraging Responsible Access to the Countryside
    Lesley Roberts and Fiona Simpson   67

6   The Host-Guest Relationship and its Implications in Rural Tourism
    Hazel Tucker   80

7   Animal Attractions, Welfare and the Rural Experience Economy
    Derek Hall, Lesley Roberts, Françoise Wemelsfelder and Marianne Farish   90

8   Authenticity – Tourist Experiences in the Norwegian Periphery
    Mette Ravn Midtgard   102

9   Rural Tourism and Film – Issues for Strategic Regional Development
     W. Glen Croy and Reid D. Walker   115

PART 4: STRATEGY AND MANAGEMENT

10   Strategy Formulation in Rural Tourism – An Integrated Approach
     Hans Embacher   137

11   Networking and Partnership Building for Rural Tourism Development
     Alenka Verbole   152

12   Integrated Quality Management in Rural Tourism
     Ray Youell   169

13   The Role of Education in the Management of Rural Tourism and Leisure
     Patricija Verbole   183

14   Ecotourism for Rural Development in the Canary Islands and the Caribbean
     Donald Macleod   194

15   Relationships Between Rural Tourism and Agrarian Restructuring in a
     Transitional Economy: The Case of Poland
     Lucyna Przezbórska   205

PART 5: CONCLUSION

16   New Directions in Rural Tourism: Local Impacts and Global Trends
     Lesley Roberts, Morag Mitchell and Derek Hall   225

Index   *235*

List of Figures

2.1   A tripartite theoretical framework of tourism impacts   22
2.2   An alternative impact model   23
2.3   The modified/extended Doxey's Irridex model   31

3.1   A model of sustainable rural tourism governance   44

9.1   *The Piano*, used in a New Zealand promotional poster   124

10.1   Increase in personal enquiries for farm holidays brochures
        to the Federal Association, 1995-2002   145

11.1   Pubs' networks in Pišece   156
11.2   The five family clans and the CIB partnership   159
11.3   Marko's family clan   160
11.4   Participation of local people in Pišece's associations,
        societies and clubs   161

12.1   The integrated quality management (IQM) process   174

15.1   Geographic distribution of agritourism and rural tourism
        enterprises in Poland, 2000   207
15.2   Number of agritourism farms per 1000 of agricultural farms
        in Poland by regions, 2000   208
15.3   Frequency of age groups of rural tourism and agritourism
        enterprises in Wielkopolska survey (2000)   211
15.4   Wielkopolska survey: entrepreneurial classification
        by age and gender   212
15.5   Surveyed farms by agricultural land classes   214
15.6   The surveyed farms by global and commodity crop production   215
15.7   The surveyed farms by global and commodity animal production   216
15.8   Surveyed rural tourism and agritourism enterprises by
        the number of income sources   218
15.9   Main and additional sources of income of the surveyed farms   219
15.10   Share of agritourism/rural tourism income in the total income
        of the interviewed households   220

List of Tables

2.1   Butler's factors of tourism-induced social change   24

3.1   Tourism and agriculture in Britain compared   50

5.1   A typology of interpretation users   74

7.1   Experiential classification of animal-based attractions   94

9.1   Studies of literary tourism   117
9.2   Cinema admissions in the USA, United Kingdom and Australia   118
9.3   Effects of rural and natural setting films on visitor numbers   122

12.1   Core principles of integrated quality management   175

15.1   Characteristics of rural areas of the Wielkopolska province
       and Poland as a whole (2001)   209
15.2   Rural area definitions applied to the Wielkopolska region and to
       Poland as a whole (2000)   210
15.3   Rural tourism and agritourism enterprises and their agricultural
       characteristics (2000)   213
15.4   Changes within farming introduced because of agritourism
       activity   217

List of Contributors

Steven Boyne, Researcher, Tourism Research Group, The Scottish Agricultural College, Auchincruive, Ayr, Scotland.

W. Glen Croy, Lecturer in Tourism Management, School of Tourism and Hospitality, Waiariki Institute of Technology, Rotorua, New Zealand.

Hans Embacher, Director of the Austrian Farm Holidays Association, member of the Marketing Council of the Austrian National Tourist Office, and board member of EuroGites.

Marianne Farish, Research Assistant, Animal Behaviour and Welfare, The Scottish Agricultural College, Bush Estate, Edinburgh, Scotland.

Derek Hall, Professor of Regional Development and Head of the Tourism Research Group, The Scottish Agricultural College, Auchincruive, Ayr, Scotland.

Donald Macleod, Head of the Tourism Research Centre, University of Glasgow, Crichton campus, Dumfries, Scotland.

Mette Ravn Midtgard, Senior Researcher, NORUT Social Science Research Ltd., Tromsø, Norway.

Morag Mitchell, Researcher, Rural Policy Research Group, The Scottish Agricultural College, Craibstone, Aberdeen, Scotland.

Per-Åke Nilsson, Senior Lecturer, Mid-Sweden University, Östersund, and affiliated to the Centre for Regional and Tourism Research, Nexø, Bornholm, Denmark.

Lucyna Przezbórska, Researcher and Lecturer, Department of Agri-Food Economics, Agricultural University of Poznan, Poland.

Lesley Roberts, Senior Lecturer, Centre for Travel and Tourism, University of Northumbria, Newcastle-upon-Tyne, England.

Richard Sharpley, Reader in Tourism Management, Centre for Travel and Tourism, University of Northumbria, Newcastle-upon-Tyne, England.

Fiona Simpson, Environmental Planner and an Associate of Land Use Consultants, Glasgow, Scotland.

Anders Sørensen, Freelance researcher, previously affiliated to the Centre for Regional and Tourism Research, Nexø, Bornholm, Denmark.

Hazel Tucker, Lecturer, Department of Tourism, University of Otago, Dunedin, New Zealand.

Alenka Verbole, Recently finished post-doctoral studies on rural development in the newly emerging context of EU, at the Agricultural University, Athens, Greece.

Patricija Verbole, Owner/manager of Le Clerck Consulting for Education and Tourism, Kamnik, Slovenia, and Lecturer in Tourism and Hospitality Programmes at the Adult Education Centre, Ljubljana.

Reid D. Walker, Masters Graduate, Department of Tourism, University of Otago, Dunedin, New Zealand, now working in sales and marketing.

Françoise Wemelsfelder, Senior Research Scientist, Animal Behaviour and Welfare, The Scottish Agricultural College, Bush Estate, Edinburgh, Scotland.

Ray Youell, Head of Tourism Management, University of Wales at Aberystwyth.

Preface

The origins of this volume lie in the international conference *New Directions in Rural Tourism and Leisure: Local Impacts, Global Trends*, hosted by the Scottish Agricultural College (SAC)'s Leisure and Tourism Management Department at its Auchincruive, Ayrshire campus, in September 2001. The book comprises substantially revised and updated selected contributions from that conference, supplemented by additional invited contributions as well as introductory and concluding chapters by the editors emphasising the key themes and coherence of the overall work.

A volume entitled *'New Directions...'* by definition is set within the context of a number of dynamic processes. Rural tourism and recreation is one of the fastest growing elements of tourism, itself claiming to be one of the world's largest and most rapidly expanding sectors. Much of the globe is experiencing rural restructuring in its various forms, and 'Western' economies in particular are witnessing the emergence of a post-productivist, post-industrial countryside. Although the context and pressures for such restructuring may differ between regions and countries, the role of tourism is a potentially important element of the restructuring process. This requires a heightened awareness of the importance of, and provision of the appropriate tools for, the sustainability of tourism-related rural development. Further, as an important and growing element of many national and regional economies, rural tourism is being employed as a key component of national and regional image construction and projection, not least in relation to the conservation, interpretation and presentation of natural and built heritage. This volume reflects these trends and processes through a balance of theory, critique and practical experience, illustrated by a wide range of empirical research findings from around the world.

The book's 16 chapters are divided into five interlinked sections. In the first section, in their introductory chapter *Tourism and the Countryside: Dynamic Relationships*, the editors establish the context for, and outline some of the elements of the dynamic relationships between, tourism and rural change and development. The three chapters in Part 2 *Conceptualisation*, each take a particular approach to key conceptual dimensions of rural tourism development. In Chapter 2, Steven Boyne provides new conceptual insights with a critique of *New Directions in Rural Tourism Impact Research*. This attempts to locate tourism impacts research within the wider tradition of tourism research, evaluating some of the characteristics and shortcomings of such research and addressing methodological problems. This is well complemented by Richard Sharpley's tightly argued critique of *Rural Tourism and Sustainability* in Chapter 3. Using the events surrounding the UK's foot and mouth disease crisis in 2001, he evaluates the contested nature and value of sustainable rural tourism development. Drawing on largely Danish experience, Anders Sørensen and Per-Åke Nilsson's provocative *What is Managed When Managing Rural Tourism?* aims to further stimulate the conceptual debate on the relationships within and between rural tourism and recreation, rural development and the management of the countryside. They compare the conception of rural tourism as it appears in the research literature with qualitative data on tourism and recreation experiences in rural areas, and argue for a de-differentiation of the symbolic boundaries between tourism and recreation. These three chapters aim to provide an appropriately contested set of conceptual approaches with which to approach and appreciate the following exemplification of rural tourism development experience.

In Part 3 *Experience*, empirical material is drawn together from Western and Northern Europe and Australasia. In Chapter 5, Lesley Roberts and Fiona Simpson's *Encouraging Responsible Access to the Countryside* leads this section by evaluating the dynamic UK legislative context for access to the countryside, placing a changing empirical situation within a debate on responsibility within tourism and the role of codes of conduct. For Chapter 6, *The Host-Guest Relationship and its Implications in Rural Tourism* by Hazel Tucker presents a well argued anthropological examination, drawing on exchange theory, of the small-scale and often highly individualistic interrelationships between hosts and guests in small rural tourism accommodation businesses. This is set within the context of South Island, New Zealand. *Animal Attractions, Welfare and the Rural Experience Economy* is Chapter 7, by Derek Hall, Lesley Roberts, Françoise Wemelsfelder and Marianne Farish. It draws on research undertaken in Scotland into the role of animals in rural tourism attractions, and sets it within the context of the experience economy, raising issues relating to the role of interpretation in an era when the recreational role of animals may be perceived to be more health-threatening. Moving further north geographically, Chapter 8 by Mette Ravn Midtgard, *Authenticity – Tourist Experiences in the Norwegian Periphery,* employs a qualitative approach to reveal how independent tourists in the Lofoten archipelago express their tourist experiences in relation to concepts of 'authenticity'. The chapter highlights notions of authenticity attached to both the natural and cultural environments of the region. The final chapter (9) in this section of the volume, *Rural Tourism and Film – Issues for Strategic Regional Development* by W. Glen Croy and Reid D. Walker, addresses the role of fictional media in developing both the tourism industry and a tourism identity in rural areas. The chapter develops specifically from New Zealand experience and goes on to focus on the role of film in the generation of image and identity and their contribution to regional development. This is set within the wider context of the relationship between popular culture, rural tourism and economic development.

Part 4 *Strategy and Management* opens with a practitioner's viewpoint, a chapter which explicitly does not follow the usual academic norms, in presenting Hans Embacher's *Strategy Formulation in Rural Tourism – An Integrated Approach* (Chapter 10). As one of the most famous and successful rural tourism practitioners in Europe, Hans' contribution on the strategy development process, drawing on substantial experience from Austria, is a welcome addition to the volume. Maintaining a Central European flavour, Chapter 11, *Networking and Partnership Building for Rural Tourism Development*, showcases Alenka Verbole's community-based research in Slovenia. It emphasises the importance of local networks and organisational practices in the rural tourism development process. In Chapter 12, Ray Youell's *Integrated Quality Management in Rural Tourism* explores an area which has grown rapidly in importance in recent years. This chapter illustrates the process, opportunities and challenges offered by establishing integrated quality management (IQM) in rural tourism and concludes that IQM has a pivotal role to play in improving the competitive position of the rural tourism sector while at the same time safeguarding social, cultural and environmental integrity. The following chapter (13), by Patricija Verbole, *The Role of Education in the Management of Rural Tourism and Leisure* examines the often clear need for rural tourism management to be informed by good and appropriate education and training. Communication knowledge and skills, as well as personal development skills, are seen as playing a vital role in improving the quality of customer experience, and in better self-promotion of the tourist product and services. Set within a context of the increasing importance of rural tourism in the development of peripheral communities around the world, Donald Macleod provides, in Chapter 14 – *Ecotourism for Rural Development in the Canary Islands and the Caribbean* – a strong critique of the management of ecotourism in contrasting island situations on both sides of the Atlantic. He emphasises the local and global pressures for ecotourism to commercialise, and reinforces the need to understand the host culture within which ecotourism takes place. The final chapter (15) of this section of the volume focuses on post-communist processes in Eastern Europe. Lucyna Przezbórska's *Relationships Between Rural Tourism and Agrarian Restructuring in a Transitional Economy: The Case of Poland*, sets the management and development of rural tourism firmly within the dynamic context of rural restructuring under conditions of national and regional political and economic change. In particular, agritourism is seen as a relatively new and untried concept for Poles, although it is perceived, rightly or wrongly, as an easier route to rural diversification and restructuring than most available alternatives.

In Part 5, the book concludes with the editors, in Chapter 16 *New Directions in Rural Tourism: Local Impacts and Global Trends*, drawing together the key themes of the previous chapters and projecting a forward agenda for rural tourism research and development. This emphasises the need for the academic and practitioner elements of management, research and development to better exploit the synergies of collaboration.

Acknowledgements

Whilst looking forward, this volume also looks back to the very recent past. It is dedicated to the short but fruitful life of SAC's Leisure and Tourism Management Department. This Department, which hosted the conference from which this volume has evolved as well as a previous international gathering (Hall and

O'Hanlon, 1998), became an unwitting casualty of serial institutional restructuring consequent upon the expert advice of a high-profile management consultancy. In this case, all those who contributed to the Department's development and strength are not too many to be mentioned:

James Adams, Ronnie Ballantyne, Alison Beeho/Mackintosh, Moira Birtwistle, Steven Boyne, Mike Burr, Fiona Carswell, Andrew Copus, Rachel Darling, Chris Doyle, Roger Evans, Isabelle Frochot, Claire Gallagher, Joy Gladstone, Deborah Gourlay, David Grant, Jacqui Greener, David Hume, Jim Kean, Richard Kelly, Irene Kirkpatrick, Lennox Lindsay, Yvonne Loughrey, Marsaili MacLeod, Pamela Marr, Joanne McDowell, Stephen Miles, Linsey O'Hanlon/Hunter, Scott Petrie, Robert Rawlings, Fiona Simpson, Stephen Smith, Nick Tzamarias, Fiona Williams, and of course, ourselves, the editors.

Warmest thanks are due to all participants and supporters of our two international conferences, to all the contributors to this volume, to our colleagues at Ashgate who have facilitated this publication, but most especially to Irene Kirkpatrick whose organisational, administrative and secretarial skills have contributed in large measure both to the success of the conference upon which it is partly based and to the co-ordination and production of this book in its camera-ready form.

Derek Hall, Lesley Roberts, Morag Mitchell
Maidens, Jesmond Vale and Insch

Reference

Hall, D. and O'Hanlon, L. (eds) (1998), *Rural Tourism Management: Sustainable Options*, The Scottish Agricultural College, Auchincruive.

List of Abbreviations

B&B	bed and breakfast
bn	billion
BPSG	Bute Partnership Steering Group
BTA	British Tourism Authority
CA	Countryside Agency (England and Wales)
CD-ROM	compact disk, read-only memory
CEE	Central and Eastern Europe
CIB	CRPOV Initiative Board: group encouraging rural dynamism and opportunities for young people (Slovenia)
CMSC	Culture, Media and Sport Committee (UK)
CRE	Centre for Rural Economy, University of Newcastle-upon-Tyne
CRPOV	National Programme for Rural Development (Slovenia)
CTO	Cyprus Tourism Organisation
DCMS	Department for Culture, Media and Sport (England)
DG	Directorate General (European Commission)
DNP	National Parks Directorate (Dominican Republic)
DRC	Danish Research Council
€	Euro
EAAE	European Association of Agricultural Economists
EC	European Commission/European Community
EMF	European Mountain Forum
ERDF	European Regional Development Fund
Erleben	a state in which the individual and perceived surroundings melt together
ESF	European Social Fund
ETC	English Tourism Council
EU	European Union
EuroGites	European Federation for Farm and Village Tourism
FAO	Food and Agriculture Organisation of the United Nations
FEDER	EU Structural Development Fund
FIT	free independent traveller
FMD	foot and mouth disease
FTE	full-time equivalent
GATT	General Agreement on Tariffs and Trade
GDP	gross domestic product
GNP	gross national product
GUS	Central Statistical Office of Poland
ha	hectare
ICIMOD	International Centre for Integrated Mountain Development

IFAW	International Fund for Animal Welfare
INTERREG	European Community initiative to stimulate interregional co-operation
IP	involvement profile
IQM	integrated quality management
IRS	Institute of Rural Studies, University of Wales at Aberystwyth
IT	information technology
LEADER	EU rural development programme (Liaison Entre Actions de Développement de l'Économie Rurale)
mn	million
NCWOR	non-consumptive wildlife-oriented recreation
NGO	non-governmental organisation
NZPA	New Zealand Press Association
OECD	Organisation for Economic Co-operation and Development
PCV	(US) Peace Corps volunteer
PHARE	Poland/Hungary Assistance for Economic Reconstruction (EU policy extended to much of the rest of CEE)
PPE	post-production exposure
RDC	Rural Development Commission (England and Wales)
rorbu	fishermen's cabins (Norway)
RTO	regional tourism organisation (New Zealand)
SAC	The Scottish Agricultural College
SEE	South-eastern Europe
SEERAD	Scottish Executive Environment and Rural Affairs Department
sme	small and/or medium sized enterprise
smte	small and/or medium sized tourism enterprise
SNH	Scottish Natural Heritage
SPADA	Screen Producers and Directors Association of New Zealand
TIAS	tourism impact attitude scale
TNC	The Nature Conservancy (Dominican Republic)
TØI	Transportøkonomisk institutt, Oslo
TQM	total quality management
TV	television
TZS	Slovenian Tourist Association
UK	United Kingdom of Great Britain and Northern Ireland
UNCED	United Nations Conference on Environment and Development
UNEP/IE	United Nations Environment Programme, Industry and Environment
UNESCO	United Nations Education and Science Organisation
USA	United States of America
USAID	United States Overseas Aid Programme
VF	visiting friends
VFR	visiting friends and relatives
VR	visiting relatives
VS	VisitScotland (Scottish tourist board)

Wahrnehmung	evocative mood
WCED	World Commission on Environment and Development
WHS	World Heritage Site
WTO	World Tourism Organisation
WTTC	World Travel and Tourism Council
www	world-wide web

*This volume is dedicated to the short but fruitful life of the
Scottish Agricultural College's Leisure and Tourism Management Department
at Auchincruive and Craibstone, 1996-2003*

PART 1
CONTEXT

Chapter 1

Tourism and the Countryside: Dynamic Relationships

Derek Hall, Morag Mitchell and Lesley Roberts

Introduction

Rural tourism development attracted increasing interest in the 1990s and a growing literature has contributed to our understanding of it as an evolving phenomenon. According to Long and Lane (2000), rural tourism has moved into its second phase of development, its first having been characterised by growth in participation, product and business development, and partnership. Its second is predicted to be more complex, and is likely to be, given the questions that remain regarding its place in policy, its integration in practice, and its dynamic role within the restructuring countryside and within wider tourism development processes. This introduction aims to establish a context for the subsequent and more focused chapters that follow by outlining issues relating to, dimensions of, and questions surrounding the dynamic nature of rural tourism and recreation.

Tourism and Recreation in Rural Areas: Key Issues

Estimates often suggest that tourism in rural areas may make up 10-25 per cent of all forms of tourism activity (e.g. EuroBarometer, 1998). But a long recognised analytical constraint is the absence of systematic statistical sources for 'rural tourism' (Lane, 1994). This is not surprising given that there may be no difference either in terms of location or activity, between 'rural tourism' and 'countryside recreation'. Many rural tourists and recreationalists are excursionists (day visitors) rather than those making overnight stays (the extent of whom can to some extent be measured in terms of bed-nights). Rural tourism's very diversity and fragmentation sees tens of thousands of enterprises and public initiatives active across Europe, some of which are listed with local or regional bodies such as tourism boards and authorities, while others are not. Overarching these issues are often fundamental differences in national definition and enumeration: one country may include only farm and nature dimensions in its conception of 'rural tourism', while another will consider many economic activities located outside of urban areas.

In many parts of the world, rural areas have long provided the setting for recreation and tourism activities, which have not always been explicitly considered or branded as 'rural'. In recent decades, however, there has been a greater industry awareness of a requirement to segment and brand various aspects of tourism and recreation just at a time when the relationships between such activities and their rural contexts have been changing and becoming more complex. Such complexity and change in rural sector relationships reflect both the dynamic and often uncertain economic and social environment in which rural development processes take place; and the growing global importance and diversity of tourism and recreation activities, with the pressures and inter-linkages (global-local, urban-rural) which that brings.

Most notably, recreation and tourism activities in a number of rural settings have been dramatically transformed from being relatively passive and minor elements in the landscape to become active and significant agents of environmental, economic and social change. Such changes have attracted attention from local, regional, national and supranational policy makers. However, this is not to suggest any consistent approach to, or agreement upon the nature, development and significance of tourism and recreation in rural areas.

What is clear is that a number of key demand factors have raised interest in rural areas, larger numbers of people are visiting rural areas, and the recreational activities undertaken in rural areas are increasing and diversifying, raising issues of competing traditional/new, passive/active pursuits and the need for adequate planning and management to cope with contrasting demands from mass and niche requirements. In the face of homogenised globalism and what may be increasingly impersonal and unsafe urban environments, rural tourism is often perceived as able to meet growing demands for personal contact, individualism, authenticity and heritage, said to reflect increasing levels of education, health consciousness, and the development of accessible high performance outdoor equipment (Long and Lane, 2000). Coupled to improved transport and communication technologies – which themselves have rendered true remoteness a rare quality and an almost unique selling point – and the residential demand for access to rural areas from both working and retired people, tourism and recreation have become an important part of a range of opportunities variously holding out the promise of economic, social, cultural and environmental enhancement.

Tourism is widely perceived as being of considerable economic and social benefit to rural areas through the income and infrastructural developments it may bring particularly to marginal and less economically developed regions. It can provide organic, relatively low capital, economic growth for locally owned business and offers a potential alternative both to traditional rural activities and to rural workers themselves (Bollman and Bryden, 1997; Long and Lane, 2000). It can of course also stimulate in-migration and the attraction of urban-based entrepreneurs who may merely siphon off any benefits away from the local rural area. Nonetheless, considerable attention has been given in the European Union to the support and enhancement of rural tourism initiatives (Mormont, 1987; Bethemont, 1994; Nitsch and der Straaten, 1995; Hjalager, 1996; Priestley *et al.*,

1996), within the wider context of rural development.   But, national and supranational organisations, government views and industry perceptions may differ or even conflict.   Notably, industry vested interests may result in over-inflated expectations for rural tourism development such as the World Tourism Organisation's claim of 'Rural tourism to the rescue of Europe's countryside' (WTO, 1996; Butler *et al.*, 1998).   Unmet expectations can easily lead not only to disappointment but to disillusionment and may actually accelerate processes of economic decline and out-migration.

Impacts of Tourism and Recreation in Rural Areas

Many parts of Europe have experienced a century, and North America some eighty years, of rural decline (Long and Lane, 2000). While an economic revitalisation of rural areas is often sought, few rural dwellers, either new or recent in-migrants, would wish to change dramatically the physical character and ethos of their landscapes by encouraging the siting of a gambling casino, prison or nuclear power station. While those activities may appear extreme, tourism, often viewed by many rural regions as one of the few opportunities to enhance the local economy, may have equally profound impacts on its countryside contexts. The challenges of rural restructuring, the major potential threats to rural environments and the dynamic social composition of many rural areas, require an understanding and management of rural tourism which is firmly integrated into an appreciation of the (often urban-derived) dynamic social, economic, political, cultural, psychological and environmental processes shaping both reality and our social construction of 'the rural'.

In contributing to successful rural development tourism needs to be employed as part of a portfolio of strategies. Tourism and recreation is not an appropriate development tool for all rural areas, but factors of comparative advantage will vary considerably from one type of rural area to another. Rural tourism is usually best suited to act as a complement to an existing thriving and diverse rural economy: within an already weak economy it can create income and employment inequalities if not complemented with other employment generating development processes Butler and Clark, 1992). Further, a number of factors can reduce its economic effectiveness. These include income leakages, volatility, a declining multiplier, low pay, imported labour, the limited number of entrepreneurs in rural areas, and the conservatism of rural investors (Lane, 1994).

Evaluation of rural tourism's development impacts raises such questions as what are the trade-offs between the social and environmental (negative) impacts and economic benefits? How far, for example, do rural designations of special area status designed to protect environments actually act to attract and focus tourists in self-fulfilling honeypots? Can the benefits to one sector (e.g. nature-based tourism assisting species conservation) be realistically measured against the negative impacts felt in a related sector (e.g. nature-based tourists' actual disturbance to wildlife)? Most tourists in rural areas are urbanites, so who benefits and who loses from the development of tourism and recreation in rural areas? Certainly the debate over the mutual misunderstanding between town and country has seen increasing politicisation of the perceived conflict between urban and rural values and aspirations. How far are such benefits and losses set within wider social and economic processes of unequal access to resources and opportunities? To what extent do our own roles, values and vested interests influence our perceptions of whether impacts are positive or negative?

The dynamic relationship between tourism and recreation and other aspects of the rural economy, society and environment is a fundamental underlying theme of this book. In summary, the contribution of tourism to rural development can include:

- revitalising and reorganising local economies, and improving the quality of life;
- supplementary income for farming, craft and service sectors, although most types of diversification render a relatively small contribution to average farm business income (e.g. McNally, 2001);
- opening up the possibility of new social contacts, especially in breaking down the isolation of remoter areas and social groups (Gladstone and Morris, 1998, 1999);
- providing opportunities to re-evaluate heritage and its symbols, 'natural' resources of landscape and the accessibility of open space, and the identity of rural places;
- assisting polices of environmental, economic and social sustainability; and
- helping to realise the economic value of specific, quality based production of foodstuffs, as well as of unused and abandoned buildings, unique scenery, spaces and culture.

Taking the last point, a growing interest in speciality foods, regional gastronomy, healthier eating and the promotion of local identity has seen the growth of food tourism as an important element in embedding rural tourism within local economic back-linkages while reinforcing a local quality image (e.g. Ilbery and Kneafsey, 1998, 1999; Brunori and Rossi, 2000; Murdoch *et al.*, 2000; Sage, 2003). However, Winter (2003) argues that the turn to local food may encompass several different forms of agriculture, contrasting rural economic contexts and a variety of consumer motivations.

But tourism can be a relatively fragile element of rural development:

- inward investment, new firm creation and employment generation may be limited owing to the small scale and dispersed nature of the industry which tends to offer low returns on investment;
- it requires many skills to be successful;

- it tends to be in the hands often of those rural entrepreneurs, such as farmers, small town and village business and local officials, who often do not have specific training in tourism;
- it involves many micro-enterprises;
- capital is often in short supply; and
- the time scale for success is usually short (Cavaco, 1995; Lane, 1998).

Rural Change and Restructuring

Restructuring processes have been apparent across most industrialised countries since at least the 1970s, and rural areas have not been exempt from significant economic, social and political change. Prior to the Second World War, the rural systems of most developed countries retained a degree of homogeneity and distinctiveness, despite the growth of commercial agriculture. This is often no longer the case as the weakening of former structures has been brought about as the result of several dimensions of restructuring:

- delayed reform of inconsistent protectionist subsidy systems such as the EU's Common Agricultural Policy (CAP), which have distorted the structural and spatial dimensions of agricultural production (Jenkins *et al.*, 1998);
- the inability of many marginal areas to shift to a more capital intensive economy (Brown and Hall, 2000);
- the selective industrialisation of much of the remaining agricultural sector;
- the pressures of urban and ex-urban development (Butler and Hall, 1998a, 1998b); and
- political and economic transformation in post-communist Central and Eastern Europe raising policy issues contrasting to, yet interrelated with, those of Western Europe, in the latter case in terms of both the adoption of Western 'advice' and models, and the aspiration for integration through EU accession (Hall and Danta, 2000; Hall, 2004).

Factors which have been responsible for profound changes in agriculture and for the people who depend on it, have also contributed to rural areas' attractiveness for many (ex-urbanites) to live and work. Mechanisation has drastically reduced agrarian labour requirements, stimulated continuing rural to urban or rural to rural migration, and has rendered both residential and non-residential properties available for new uses or for the same uses by different residents with often markedly different values and life-styles. At the same time, the trend towards a greater centralisation of governmental and commercial activities has contributed to a reduction or elimination of much service provision in rural settlements.

The combination of these two factors can result in villages no longer able to function, as decreasing numbers of farms and diminishing agrarian populations reduce the labour force and weaken the local community's ability to sustain the previous range of goods and services. In certain rural areas, particularly those where agribusiness does not have a dominant presence, rural repopulation by non-farm populations may take place. Stimulated by a demand for primary or second homes suitable for retirement, commuting and leisure purposes, such repopulation has contributed to notable demographic and socio-economic change in a number of rural regions.

The level of social integration resulting from these processes may be crucial for the health and self-perception (and thus the 'imaging') of rural areas. New, ex-urban inhabitants of rural towns and villages may attempt to conserve a rural idyll while opposing any form of modern development that is likely to spoil their new-found rural life-style. And villages may be in danger of becoming fossilised and sterilised as a result. The gentrification of erstwhile agricultural areas may act to convert villages into little more than extensions of bourgeois suburbs with complementary luxury and fashion shops acting to force out more basic service functions. In the Netherlands, for example, it is claimed that the agricultural productivist image of the countryside has been replaced by a more consumption-oriented and idyllic image (Huigen, 1998; de Haan, 2001; van Dam *et al.*, 2002).

For tourism, the implications of this may seem contradictory. 'Typical' rural villages are likely to attract visitors eager to experience the image of rurality, created, directly or indirectly, by counter-urbanisation. Yet tourists may not be welcomed by the rural newcomers who may have been instrumental in 'preserving' or reinforcing that landscape with such images.

Further factors important in stimulating and sustaining counter-urbanisation have included businesses relocating into rural or peri-urban areas to take advantage of lower costs and ease of transport access via regional highways. Ironically, incomers may have an important role to play here. The positive EU view is that such newcomers may bring new entrepreneurial skills and an invigoration of the local demographic profile (e.g. Anon, 2000: 18-22). Their numbers may also be sufficient to help a local rural area retain or even revitalise services and facilities such as schools, post offices and convenience goods shops.

The intensity of such processes is variable both within and between countries, as it is dependent upon a combination of local, regional and national factors. But one adverse side-effect of this 'counter-urbanisation' is the inflation of rural house prices, compounding a longer-term trend for young people born in the countryside to migrate to urban areas for perceived better employment and life-style opportunities.

The interrelated rise of agri-business and decline of traditional family farming in many areas have brought striking changes to the nature and function of farms and agrarian landscapes. The social and economic influence of both large corporate and small individual ex-urban property owners has seen changes in attitudes towards both the control of and access to rural land. As a result, as demands become more complex and varied, and are increasing rapidly, in many rural areas the resources and opportunities to meet these demands are also changing and may be diminishing, relatively if not absolutely.

Forces of rural change can be viewed as deriving from two types of sources. Endogenous forces include reduced protectionism, policies supporting multiculturalism, population loss – especially of younger skilled people – or gain, ageing populations, increased leisure time, changing family structures. Exogenous forces may relate to the operation of transnational corporations, technological innovation, global financial markets and economic restructuring. Certainly rural restructuring is enmeshed in the process and influences of globalisation, and cannot be viewed in isolation from it.

The effects of rural change may be masked or overlain within many near-urban or fringe zones which remain 'rural' in administrative, and thus also in statistical, terms. Data for economic activity, unemployment and population decline in rural areas may be severely distorted in these areas by such factors as the influx of retired newcomers, the inclusion of country dwellers who commute to work in urban areas, and by part-time and seasonal employment. Indeed, 'traditional' rural dwellers may feel that they are losing their identity and/or are being marginalised by such developments. Inconsistency and paradox in patterns of rural socio-economic role and status may arise here due to the extremes in wealth and life-styles within and between different types of rural dwellers and resulting from local and regional contrasts in the migration component and characteristics of demographic change.

One of the major infrastructural changes resulting from recreational development in rural areas has been a growth in second home ownership, especially in more popular destination areas, a phenomenon which may result in tourism being blamed for local housing problems. But while leisure development may indeed contribute to these, it may be just one of a matrix of factors, and in some rural areas may play no part at all. Planning policies, restraints on housing development, changes in market trends for and availability of rented property, as well as patterns of demographic and economic change, may all be more significant than the impact of recreation-related factors. However, in some popular tourist regions, establishing an acceptable and sustainable balance between demand pressures from incomers for the purchase of rural residences, and the often pressing housing and associated needs of 'traditional' rural residents, represents a major policy challenge, and can lead to the demand for total exclusion of further second home ownership in particular locations.

Changing Recreational Demands on Rural Resources

Until perhaps thirty years ago most leisure activities in rural areas were related closely to the intrinsic environmental setting, and could be characterised as relaxing, relatively passive, perhaps nostalgia-related, with forms of activity such as walking, picnicking, fishing and landscape photography. Many of these activities represented escape from urban life into an environment of contrasting pace and setting where the physical and human elements were thought to blend in harmony. While such 'traditional' recreational practices are still widely pursued in rural areas, other activities have grown in appeal. These are generally more active, competitive, prestige- or fashion-related, perhaps technological, modern, individual and fast, such as survival games, off-road motor vehicle driving and hang gliding (Butler, 1998; Butler *et al.*, 1998). These may be viewed as acting to impose urban values on rural areas, while the specific locational context may be far less important and in some cases almost irrelevant as the activities themselves are not focused on intrinsic rural qualities. Greater need for the penetration of remote areas by motorised transport and a demand for specific facilities and resorts also result from the sophisticated and often more impacting demands placed upon rural resources by these activities.

Global forces and fashions have contributed to a rising demand for activities such as golf and other sports, holiday and amusement parks which are not 'rural' in nature or scale but tend to be located in rural areas (Butler, 1998). Such activities may be present as the result of product-led external development forces (such as the location of a *Center Parcs* complex) or may result from the demand-led expansion of a locally owned enterprise (Getz and Page, 1997).

Social Construction of Rurality

As modern (urban) life has become faster, more stressful and less 'authentic', so the symbolic significance of the countryside has taken on a more utopian, mythical role as a simpler, slower, more natural, more meaningful and a thus 'superior' state compared to the urban. Such construction may vary from individual to individual and collectively may represent cultural differences between regions and countries. Thus in England it may be a 'romantic, pre-industrial idyll, a "chocolate box" countryside of meadows, villages and country lanes' (Sharpley and Sharpley, 1997: 16). For North Americans, by contrast, recreational rurality may be conceived often as preserved 'wilderness' as typified by western national parks.

Many countries and regions are now sold to tourists through the reinforcement of such (claimed) popularly held images based on an 'authentic' rurality which is perceived (usually by tourism industry decision makers) to represent the social constructions of visitors but which may contrast and even conflict with the reality of the destination area: local residents may not concur with or may even disdain the images they are supposed to represent or be associated with. Yet such consumption and commodification of the countryside can result in rural production being focused on establishing new commodities or on re-imaging and rediscovering places for recreation and tourism. Thus with rural 'communities' becoming objects of tourism consumption, some localities have been encouraged to reproduce themselves specifically for tourists: to identify themselves with the way in which they are 'named' and 'framed' as tourist attractions. Tourism industry demands for 'authentic' local cultures which may be associated with a specific location can create 'back-stage' and 'front-stage' areas, with the tourist base being restricted to the 'staged' authenticity of the 'front-stage' regions. However, in the process, local community relations themselves may become commodified (Hall and Richards, 2003).

But while a 'post-fordist' shift has been identified empirically by such researchers as Hummelbrunner and Miglbauer (1994) and Boissevain (1996), suggesting that the balance of demand for tourist services is moving from a pattern of mass consumption to more individual patterns, with greater differentiation and volatility, there is in fact a continued global growth of mass tourism markets as successive developing countries become tourism source areas. This raises such questions as how such rural areas will be able to respond and cope if specialist interest is merely the precursor of, or 'complement' to mass demand?

Coincident with structural change, rural areas have become subject to a much greater range of functional uses. These can raise conflicts both between recreation and tourist uses and other forms of land use, and between various forms of recreation and tourism themselves: motorised and non-motorised, hikers, hunters and paintballers, pony trekkers and mountain bikers, canoeists and anglers. Such conflicts are likely to intensify as both the overall demand from recreational and tourist uses and the breadth of those uses increase (Butler *et al.,* 1998). In this use or reuse of rural areas, while a sense of place – through marketing and image promotion – may have been reasserted, a loss of spirit of place (e.g. Durrell, 1969) may have been brought about as a result of a rapidly changing and diminishing base of 'traditional' rural activities and people, as the holistic raison d'être of a rural place is all but lost. Nonetheless, it has been argued by Wilkinson and others that an ethic of place – a shared community value – should persist which respects equally the people of a region, the land, animals, vegetation, water and air (Udall, 1990).

Kotler *et al.* (1993: 100) have considered place in terms of four components: character, fixed environment, service provision, and entertainment and recreation. Yet transcending all of these is purpose: the basis for the development of that place within its particular environmental context and with its specific demographic and cultural characteristics. Certainly the importance of image for rural areas has taken on a new dimension with the adoption of IT for marketing and place promotion (e.g. Boyne *et al.*, 2000, 2003). Major efforts are often made in a variety of rural settings to consciously 'improve', establish or change a sense of place through the creation and recreation of specific images. The diverse range of purposes of such images raises interesting research questions concerning their selection processes and the balance of interests represented within them. Walmsley (2003) has recently argued that while there is often an element of blind optimism in the reliance of rural tourism as some sort of economic panacea, as life-style becomes an increasing determinant of recreational activity, rural communities will be able to capitalise on this through appropriate place marketing.

Images of rural areas would appear to be largely 'positive' in most countries of the developed world (e.g. van Dam *et al.*, 2002), because of, or perhaps despite, the images which are promoted for residential, investment and tourism purposes. 'Strategic place marketing' (Kotler *et al.*, 1993) is seen to be necessary to meet the perceived needs of a locality's stakeholders in order to 'improve livability, investibility and visitability' (Kotler *et al.*, 1993: 99), and involves the construction or selective tailoring of particular images to project within the global market place.

Being able to capture, articulate and represent local 'community' life and its relationship with both the perceived and hidden landscape may be an essential element of the branding and image projection of the rural tourism product, and it would appear logical to avoid damaging or compromising the way of life around which local attraction is based (Edwards *et al.*, 2000). Yet the generation, commodification and impact of 'idyllic' images of timeless sustainability may stifle the articulation of actual local identity or identities. The impact of rural tourism on, and the representation of gender within this context is contested. Arguments that tourism development opportunities offer rural women routes to managerial roles and positions of independence are countered by observations that, for example, running farm-based bed and breakfast enterprises is little more than an extension of women's 'traditional' domestic role. The latter activity – often undertaken by farmers' wives – may even be interpreted as acting to reinforce the patriarchal ideologies which exclude and marginalise women from farming (Saugeres, 2002).

Conclusions

In their review of rural tourism development, Long and Lane (2000) argue that rural tourism – at least in Europe and North America – is entering a more complex phase of expansion, differentiation, consolidation and understanding, and that a number of implications flow from this:

- markets will continues to develop, competition will increase and provision will grow and products further diversify, necessitating the further development of partnerships and networks in the supply and promotion of rural tourism attractions;
- more national and regional rural tourism policies will be formulated, although the nature and balance of public or semi-public intervention will vary in time and space;
- marketing and training will (need to) become more effective and sophisticated, not least through the use of IT for website promotion and information, on-line booking, customer care and distance delivery of education and training provision;
- as a consequence of increasing activity and competition, land-use issues, not least those of access to rural land, will intensify;
- heritage, and the contested power relations behind its reproduction, promotion and interpretation, will consolidate its position as an anchor of rural tourism; and
- more emphasis will be placed on the sustainability of rural tourism policy.

Long and Lane (2000) suggest that accepted indicators are needed with which to monitor the impacts of such developmental trends over a period of time, and that the use of such indicators can assist better policy making and planning.

An increasingly complex and diverse pattern of rural depopulation and counter-urbanisation, upon which tourism and recreation undertaken in rural areas largely by urban dwellers has been superimposed, has brought together, often in sharp conflict, aspirations of modernisation alongside the seeking out of a past, idealised idyll. Subjective, emotional attachments to imagined place and/or community may bear no relation to objective reality yet may influence the ways in which people behave. Representations of rurality, through tourism, residential and investment promotion may actively structure rural spaces. The demand for pretty villages or the construction and commodification of cultural associations act to shape the appearance of rural settlements in order to satisfy the needs of visitors. Awareness of such preferences needs to be articulated within development processes.

As an introduction to the dynamic nature of relationships within and between rural areas and tourism, this chapter has attempted to provide a brief:

- description of some of the key themes of rural tourism development;
- discussion of the dynamic rural social and economic context within which demands for recreation and tourism are taking place;
- assessment of the changing nature of recreation and tourism in rural areas and the potential conflicts which may arise; and
- evaluation of the social construction and commodification that rural areas may experience as part of the development processes embracing residential development and wider inward investment as well as tourism and recreation.

References

Anon (2000), 'Neo-ruralites bring relief', *LEADER Magazine*, vol. 22, pp. 18-22 <http://www.rural-europe.aeidl.be/rural-en/biblio/pop/contents.htm>.

Bethemont, J. (ed.) (1994), *L'Avenir des Paysages Ruraux Européens*, Laboratoire de Géographie Rhodanienne, Lyon.

Boissevain, J. (ed.) (1996), *Coping with Tourists*, Berghahn, Oxford.

Bollman, R.D. and Bryden, J.M. (eds) (1997), *Rural Employment: An International Perspective*, CAB International, Wallingford.

Boyne, S., Hall, D. and Gallagher, C. (2000), 'The Fall and Rise of Peripherality: Tourism and Restructuring on Bute', in F. Brown and D. Hall (eds), *Tourism in Peripheral Areas*, Channel View Publications, Clevedon, pp. 101-13.

Boyne, S., Williams, F. and Hall, D. (2003), 'Policy, Support and Promotion for Food-related Tourism Initiatives: A Marketing Approach to Regional Development', *Journal of Travel and Tourism Marketing,* vol. 14.

Brown, F. and Hall, D. (2000), 'The Paradox of Peripherality', in F. Brown and D. Hall (eds), *Tourism in Peripheral Areas,* Channel View Publications, Clevedon, pp. 1-6.

Brunori, G. and Rossi, A. (2000), 'Synergy and Coherence Through Collective Action: Some Insights from Wine Routes in Tuscany', *Sociologia Ruralis*, vol. 40, pp. 409-23.

Butler, R. (1998), 'Rural Recreation and Tourism', in B. Ilbery (ed.), *The Geography of Rural Change*, Addison Wesley Longman, Harlow, pp. 211-32.

Butler, R. and Clark, G. (1992), 'Tourism in Rural Areas: Canada and United Kingdom', in I. Bowler, C. Bryant and M. Nellis (eds), *Contemporary Rural Systems in Transition. Economy and Society*, CAB International, Wallingford, vol. 2, pp. 166-86.

Butler, R.W. and Hall, C.M. (1998a), 'Conclusion: The Sustainability of Tourism and Recreation in Rural Areas', in R.W. Butler, C.M. Hall and J. Jenkins (eds), *Tourism and Recreation in Rural Areas*, John Wiley & Sons, Chichester and New York, pp. 249-58.

Butler, R.W. and Hall, C.M. (1998b), 'Tourism and Recreation in Rural Areas: Myth and Reality', in D. Hall and L. O'Hanlon (eds), *Rural Tourism Management: Sustainable Options*, The Scottish Agricultural College, Auchincruive, pp. 97-107.

Butler, R.W., Hall, C.M. and Jenkins, J. (eds) (1998), *Tourism and Recreation in Rural Areas*, John Wiley & Sons, Chichester and New York.

Cavaco, C. (1995), 'Rural Tourism: The Creation of New Tourist Spaces', in A. Montanari and A.M. Williams (eds), *European Tourism: Regions, Spaces and Restructuring*, John Wiley & Sons, Chichester and New York, pp. 127-49.

de Haan, H. (2001), 'The Construction of Rural Futures in the Netherlands', *Tijdschrift voor Sociaalwetenschappelijk Onderzoek van de Landbouw*, vol. 16, pp. 6-23.

Durrell, L. (1969), 'Landscape and Character', in A.G. Thomas (ed.), *Spirit of Place: Letters and Essays on Travel: Lawrence Durrell*, Faber and Faber, London, pp. 156-63.

Edwards, J., Fernandes, C., Fox, J. and Vaughan, R. (2000), 'Tourism Brand Attributes of the Alto Minho, Portugal', in G. Richards and D. Hall (eds), *Tourism and Sustainable Community Development*, Routledge, London, pp. 285-96.

EuroBarometer (1998), *Facts and Figures on the Europeans' Holiday*, EuroBarometer for DG XXIII, Brussels.

Getz, D. and Page, S.J. (1997), 'Conclusions and Implications for Rural Business Development', in S.J. Page and D. Getz (eds), *The Business of Rural Tourism*, International Thomson Business Press, London, pp. 191-205.

Gladstone, J. and Morris, A. (1998), 'The Role of Farm Tourism in the Regeneration of Rural Scotland', in D.Hall and L. O'Hanlon (eds), *Rural Tourism Management: Sustainable Options*, The Scottish Agricultural College, Auchincruive, pp. 207-21.

Gladstone, J. and Morris, A. (1999), 'Farm Accommodation and Agricultural Heritage in Orkney', in F. Brown and D. Hall (eds), *Peripheral Area Tourism: Case Studies*, Research Centre of Bornholm, Naxø, pp. 111-20.

Hall, D. (ed.) (2004), *Tourism and Transition: Political, Economic and Social*, CAB International, Wallingford.

Hall, D. and Danta, D. (eds) (2000), *Europe Goes East: EU Enlargement, Diversity and Uncertainty*, The Stationery Office, London.

Hall, D. and Richards, G. (2003), *Tourism and Sustainable Community Development*, Routledge, London.

Hjalager, A-M. (1996), 'Agricultural Diversification into Tourism', *Tourism Management*, vol. 17, pp. 103-11.

Huigen, P. (1998), 'Mist, Mest en Mak', *Rooilijn*, vol. 31, pp. 266-72.

Hummelbrunner, R. and Miglbauer, E. (1994), 'Tourism Promotion and Potential in Peripheral Areas: the Austrian Case', *Journal of Sustainable Tourism*, vol. 2, pp. 41-50.

Ilbery, B. and Kneafsey, M. (1998), 'Product and Place: Promoting Quality Products and Services in the Lagging Rural Regions of the European Union', *European Urban and Regional Studies*, vol. 5, pp. 329-41.

Ilbery, B. and Kneafsey, M. (1999), 'Niche Markets and Regional Speciality Food Products in Europe: Towards a Research Agenda', *Environment and Planning A*, vol. 31, pp. 2207-22.

Jenkins, J., Hall, C.M. and Troughton, M. (1998), 'The Restructuring of Rural Economies: Rural Tourism and Recreation as a Government Response', in R. Butler, C.M. Hall and J. Jenkins (eds), *Tourism and Recreation in Rural Areas*, John Wiley & Sons, Chichester and New York, pp. 43-67.

Kotler, P., Haider, D.H. and Rein, I. (1993), *Marketing Places: Attracting Investment, Industry and Tourism to Cities, States and Nations*, The Free Press, New York.

Lane, B. (1994), 'What is Rural Tourism?', *Journal of Sustainable Tourism*, vol. 2, pp. 7-21.

Lane, B. (1998), 'Rural Tourism: Global Overviews', *Rural Tourism Management: Sustainable Options. Conference Programme*, The Scottish Agricultural College, Auchincruive, p. 3.

Long, P. and Lane. B. (2000), 'Rural Tourism Development', in W.C. Gartner and D.W. Lime (eds), *Trends in Outdoor Recreation, Leisure and Tourism*, CAB International, Wallingford, pp. 299-308.

McNally, S. (2001), 'Farm Diversification in England and Wales – What Can We Learn From the Farm Business Survey?', *Journal of Rural Studies*, vol. 17, pp. 247-57.

Mormont, M. (1987), 'Tourism and Rural Change', in M. Bouquet and M. Winter (eds), *Who From Their Labours Rest? Conflict and Practice in Rural Tourism*, Avebury, Aldershot, pp. 35-44.

Murdoch, J., Marsden, T. and Banks, J. (2000), 'Quality, Nature, and Embeddedness: Some Theoretical Considerations in the Context of the Food Sector', *Economic Geography*, vol. 76, pp. 107-25.

Nitsch, B. and der Straaten, V. (1995), 'Rural Tourism Development: Using a Sustainable Development Approach', in H. Coccossis and P. Nijkamp (eds), *Sustainable Tourism Development*, Avebury, Aldershot.

Priestley, G.K., Edwards, J.A. and Coccosis, H. (eds) (1996), *Sustainable Tourism? European Experiences*, CAB International, Wallingford.

Sage, C. (2003), 'Social Embeddedness and Relations of Regard: Alternative "Good Food" Networks in South-west Ireland', *Journal of Rural Studies*, vol. 19, pp. 46-70.

Saugeres, L. (2002), 'The Cultural Representation of the Farming Landscape: Masculinity, Power and Nature', *Journal of Rural Studies*, vol. 18, pp. 373-84.

Sharpley, R. and Sharpley, J. (1997), *Rural Tourism. An Introduction*, International Thomson Business Press, London.

Udall, S.L. (1990), *Beyond the Mythic West*, Peregrine Smith, Salt Lake City.

van Dam, F., Heins, S. and Elbersen, B.S. (2002), 'Lay Discourses of the Rural and Stated and Revealed Preferences for Rural Living. Some Evidence of the Existence of a Rural Idyll in the Netherlands', *Journal of Rural Studies*, vol. 18, pp. 461-76.

Walmsley, D.J. (2003), 'Rural Tourism: A Case of Lifestyle-led Opportunities', *Australian Geographer*, vol. 34, pp. 61-72.

Winter, M. (2003), 'Embeddedness, the New Food Economy and Defensive Localism', *Journal of Rural Studies*, vol. 19, pp. 23-32.

WTO (World Tourism Organisation) (1996), 'Rural Tourism to the Rescue of Europe's Countryside', *WTO News*, vol. 3, pp. 6-7.

PART 2
CONCEPTUALISATION

Chapter 2

New Directions in Rural Tourism Impact Research

Steven Boyne

Rationale and Aims

The post-World War II global growth of tourism has given rise to a wide range of tourism impact studies. However, with regard to tourism's social impacts (which affect human behaviour and organisation), current knowledge – which stems largely from studies of tourism in less developed countries and regions and in areas where tourism has developed within a relatively short time-frame – may not reflect accurately the conditions present in rural areas of Scotland. Owing to Scotland's relatively slow historical rate of tourism growth and the predominance of domestic tourism (annually, around 92 per cent of tourist trips in Scotland are generated from within the United Kingdom), impacts associated with rapid and unchecked tourism development, or levels of relative wealth and cultural differences between host and guest, may be less severe or non-existent in this country.

As tourism continues to be developed in rural areas in order to counter economic decline in the primary production sectors, the need for sustainable forms of development is recognised. If, however, indicators of sustainability in tourism are based on impact models which are inappropriate for use in rural Scotland, then attempts to develop tourism in a less-damaging manner may be fundamentally compromised. Accordingly, the research upon which this paper is based aims to develop a framework for understanding how residents in case study areas of rural Scotland view and respond to the social impacts of tourism. Such research is often undertaken in order to monitor the social well-being of destination areas in the presence of tourism and tourists, as the viability of an area's tourism industry can be affected negatively if deterioration is perceived to occur to the natural or social environment. Such negative perceptions on the part of areas' residents can diminish their support for tourism development and can impact upon the experience of the visitors through their interactions with them. Communities are not homogenous and can contain discrete sub-groups identifiable by their attitudes to tourism; the identification of such sub-groups can provide planning-relevant information which can aid the management of tourism development.

Studies of community reactions to tourism have been undertaken from a variety of different perspectives and have employed a diverse range of analytical techniques. However, researchers working in this area have been unable thus far to develop a unified theory which can explain variations in residents' reactions to tourism. This failure of the collective efforts of academics working in tourism impact research to find consensus has led to criticism relating to the atheoretical nature of tourism research (Ap, 1990) – mirroring Dann *et al.*'s (1988) comments relating to tourism research in general – while elsewhere, there are calls for researchers to employ more qualitative, exploratory approaches to tourism research. Although these differing approaches are not mutually exclusive (Walle, 1997: 535), satisfying them simultaneously can present researchers with certain methodological problems. Ap (1990: 615), for example, writes that for tourism research to progress, it must move from its initial descriptive phase to a more explanatory position and, elsewhere, Ap and Crompton (2001: 317) suggest that, in the context of scale development for tourism impact research, researchers should work within existing theoretical frameworks and employ deductive research approaches. While this type of approach is useful, there is a danger, however, that it can stifle the contextual sensitivity which is often required during the early stages of research, for example, work being undertaken in a new geographical context. Furthermore, empirical methods and deductive reasoning may not be the most useful approach for capturing the complex social realities which constitute tourism-related phenomena (see, for example, Walle, 1997).

Within the context described above, this paper seeks firstly to locate tourism impact research within the wider tradition of tourism research, in this way providing the reader with an overview of the development of each, and a framework for understanding the epistemological concerns relating to research approaches. Secondly, some of the characteristics and shortcomings of tourism impact research are described and the problematic methodological position referred to above is elaborated upon. Thirdly, the paper discusses Pearce *et al.*'s (1996) social representations approach to community tourism relationships and how such an approach may go some way towards bridging the 'research gap' described above. Finally, drawing upon: (i) Pearce *et al.*'s (1996) social representations approach; (ii) a review of the tourism impact and the wider tourism research literature; and (iii) insights from previous research undertaken in rural Scotland, this paper introduces a novel technique for investigation into communities' responses to tourism in rural Scotland. Specifically, the approach described employs a value-orientated focus based on residents' attitudes to 'everyday-life' and their everyday-life aspirations. In this way, it is hoped to identify discriminants of residents' reactions to tourism based on their 'everyday-life' characteristics. By employing this approach successfully, it is anticipated that a precedent can be set for future empirical research to build upon while also strengthening the underpinning theoretical frameworks in this area of tourism research.

Traditional Research Approaches

Jafari (1990) identifies four traditions, or *platforms*, within the tourism research literature and notes that, although these emerged consecutively, all remain extant.

The first of these, the *advocacy platform*, began during the post-1945 period and is typified by research which emphasises the economic and positive benefits of tourism development. Following this, during the early 1970s, the *cautionary platform* emerged and attention was drawn to the more negative consequences of tourism including environmental impacts in destination areas, problems with seasonality, pressure on local facilities and changes in hosts' lifestyles brought about by the presence of tourism and tourists. In the late 1970s Cohen (1978) argued that attention was being overly focused on tourism's negative effects – an observation which perhaps heralded the emergence of the third of Jafari's platforms, the *adaptancy platform*. This platform is characterised by the 'more balanced perspective' which Ap and Crompton ascribe to the 1980s and 1990s, '...where both positive and negative impacts are evaluated' (Ap and Crompton, 1998: 120). Additionally, during this period, several types of non-mass tourism such as 'alternative tourism' and 'ecotourism' were advocated as being less damaging to society and environment – critics, however, suggested these forms of tourism may be unrealistic and their goals, though laudable, largely unachievable (Wheeller, 1991; Wheat, 1994). Jafari's final platform, the *knowledge-based platform*, '...aims at positioning itself on scientific foundation' (Jafari, 1990: 35) and draws its theoretical base from the wider social sciences (Dann, 1996, cited in Brown, 1998: 5).

Tourism Impact Research

The development of research into the impacts of tourism can be seen as reflecting, to some degree, Jafari's platform model. Research undertaken during the 1960s often focused on the economic and positive effects of tourism (Ap and Crompton, 1998: 120), reflecting (and no doubt helping to sustain) a period of optimism about tourism's contribution to regional development objectives such as the creation of wealth, growth in GDP, infrastructure development and growth in foreign exchange earnings. During the 1970s there followed several noted works which approached tourism in a more circumspect manner, including: Young's *Tourism: Blessing or Blight* (1973); Turner and Ash's *The Golden Hordes* (1975); Finney and Watson's *A New Kind of Sugar: Tourism in the Pacific* (1977); Smith's *Hosts and Guests* (1978); and de Kadt's *Tourism: Passport to Development?* (1979). These works – undertaken largely from anthropological and sociological perspectives – were concerned with understanding how and why communities had responded to tourism and the ways in which tourism had changed peoples' way of life, alongside other forces of change and modernisation.

During the 1980s and 1990s, impact research maintained a cautionary element although this was now tempered with a (re-)recognition of tourism's more positive impacts. An increasingly large number of journal articles were published describing research which sought to establish the nature of tourism's impacts in various geographical locations (Ap and Crompton, 1998, for example, contains a reasonably detailed review of these). In contrast to the ethnographic approach seen in the literature of the 1970s, however, the research was increasingly being undertaken from a planning and development perspective and drew heavily upon hosts' perceptions as a means of identifying and measuring tourism's impacts. Additionally, researchers now began to employ social science techniques in an attempt to more accurately and rigorously identify and record tourism impacts.

Viewing tourism impact research within the context of Jafari's platform model, we can see the advocacy element in the early economic-centric work, the cautionary approach being manifested during the 1970s while the more balanced approach of the 1980s and 1990s can be related to the adaptancy platform. Additionally, the use of social science research methods such as attitude measurement scales (Davis *et al.*, 1988; Lankford and Howard, 1994; Madrigal 1995) and social science theory, for example, social exchange theory (Ap, 1992) and social representations theory (Pearce *et al.*, 1996) in much of the post-1970s work fits with the premises of the knowledge-based platform.

Tourism Impact Theory

Tourism's impacts are often considered within a tripartite theoretical framework consisting of an economic impact domain, an environmental impact domain and a social and cultural (or socio-cultural) impact domain (Figure 2.1). These domains

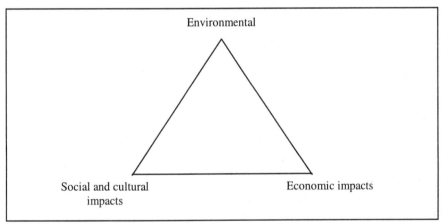

Source: after Butler (1974)

Figure 2.1   A tripartite theoretical framework of tourism impacts

are not mutually exclusive; Brougham and Butler (1981: 577), for example, position economic, environmental and cultural impacts all within one overarching domain labelled 'social impacts' (Figure 2.2). This is a supportable position: for example, an environmental impact such as the loss of a woodland (from, say, a hotel development) which residents have used as a recreational resource has clear social impacts – as would loss of employment and income have to a household or community (perhaps due to declining tourism activity in an area).

Elsewhere, Mathieson and Wall (1982: 5) acknowledge and refer to 'cross impacts', which result from the interactions between economic, environmental and social phenomena, UNESCO (1976: 75) also note the '...close interrelationship between economic and soecial (sic) aspects...' and Runyan and Wu (1979: 448) write that:

> The development of tourism can have extensive physical, social and economic impacts. Certain of these impacts can be described as relatively complex: those that take many variables to describe, are difficult to quantify and which are sensitive to policy or other difficult-to-predict interventions.

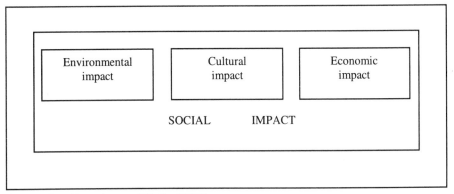

Source: from Brougham and Butler (1981)

Figure 2.2   An alternative impact model

Ap and Crompton (1998) suggest that one reason for the reporting of positive benefits in the early economic-centric studies was the relative ease of measuring variables such as income and employment, while costs such as noise, overcrowding and pollution are relatively intangible and more difficult to measure 'in economic terms' (p. 120). Perhaps this difficulty goes some way to explaining why investigations of 'complex' tourism impacts attracted ethnographic approaches as undertaken by anthropologists and sociologists during the 1970s (e.g. Smith, 1978 and de Kadt, 1979 etc, as above); it certainly is related to the substantial number of post-1970s papers dealing with 'residents' perceptions' or 'residents' attitudes' which solicit residents' opinions on tourism impacts, rather than attempting to measure these impacts directly (these papers include: Pizam, 1978; Belisle and Hoy, 1980; Sheldon and Var, 1984; Davis *et al.*, 1988; Prentice, 1993; Lankford and Howard, 1994; Madrigal, 1995; Ap and Crompton, 1998; Ryan *et al.* 1998; and Fredline and Faulkner, 2000).

The rationale for the 'residents' perceptions' approach to collecting data on the social impacts of tourism is described clearly in the Rural Development Commission's *Rural Research Report No. 21* (RDC, 1996: 4-5). This report highlights the difficulty for tourism impact researchers in establishing the *alternative position*, that is, the '...position of rural communities if there were fewer tourists, to isolate the effects of tourism' (RDC, 1996: 4). Runyan and Wu, in support of the involvement of hosts in impact research, also note that the 'involvement of residents in an impact estimation process can serve to both forecast and appraise these relatively complex impacts.' (1979: 448). Although residents have frequently been canvassed for their perceptions and attitudes on a whole range of tourism-related issues, including economic and environmental impacts of tourism, this approach has been particularly useful – due to the difficulties faced in undertaking objective measurements – in cases where relatively intangible, complex, social and cultural impacts are being measured. What, then, are these social and cultural impacts of tourism?

Tourism's Social Impacts

The model in Figure 2.1 illustrates three impact domains, economic, environmental and social and cultural (or socio-cultural – from here on referred to as social impacts): these domains are not mutually exclusive and interactions can occur as elements of complex impacts can be recognised in more than one domain (as, for example, in the woodland development scenario above). Social impacts can be described as those which affect human behaviour and organisation. Examples of these impacts include changes in residents' behaviour such as: modes of dress; sexual permissiveness; family and gender relations; lifestyle; and moral conduct. Additionally, changes may occur in the types of employment available locally, there have been cases of relocation of residents to make way for tourism developments, there can be an increase in black market activities, crime levels and the incidence of prostitution may rise and traditional cultural activities may decline (as younger people adapt to more modern leisure pursuits) or be adapted or revived for tourists' consumption. Writing in 1974, Butler proposed two basic groups of '...factors involved in determining the nature and extent of tourism-induced social change' (1974: 107). The first of these groups contained factors relating to the characteristics of the tourists and the second, factors relating to the characteristics of the destination area. These are shown in Table 2.1.

Table 2.1   Butler's factors of tourism-induced social change

Factors related to the tourists	Factors related to the destination area
1 Numbers of visitors	1 Economic state of the area
2 Length of stay of visitors	2 Degree of local involvement in tourism
3 Ethnic characteristics of visitors	3 Spatial characteristics of the tourism
4 Economic characteristics of visitors	development
5 Activities of visitors	4 Viability of the local culture
	5 Other characteristics

Source: Butler (1980)

Brunt and Courtney (1999: 495-7) posit three sub-domains of social impacts: the first concerns social impacts of tourism development (infrastructure development and employment creation, for example); the second relates to social impacts from the tourist-host interaction; and the third concerns specifically the cultural impacts of tourism. They suggest that while the first two sub-domains are active in a shorter time-frame, the cultural impacts '...lead to a longer-term, gradual change in a society's values, beliefs, and cultural practices' (pp.495-6).

This short review of tourism's social impacts is intended to provide an introduction to this area of research – for more detailed enumerations of research into tourism's social impacts the reader is directed towards, in the first instance, Pearce *et al.* (1996: 1-29) and also Brunt and Courtney (1999) and Brown and Giles (1994). For further detailed information on these impacts, the following sources offer useful analyses: Mathieson and Wall (1982); Dogan (1989); Brown (1998); and Williams (1998). Having briefly introduced the historical context for tourism impact research the paper now returns to the residents' perceptions research referred to above. The rationale behind this tradition is described, following which some of the criticisms and limitations of this research are discussed. The paper then goes on to elaborate the concerns surrounding epistemological approaches to residents' perceptions research before introducing a new approach which it is hoped can satisfy the need for exploratory research in new geographical contexts while working within existing theoretical frameworks.

Residents' Perceptions Research

To undertake the research which generated this paper, this author adopted the approach that, for an holistic analysis, the conceptual framework illustrated in Brougham and Butler (1981: 577) and shown above in Figure 1.2 is the most useful as it recognises the interlinkages between cultural, economic and environmental impacts and views all of these as *social impacts*. Indeed, in the wider tourism impact and residents' perceptions literature, impacts from all three conceptual domains (Figure 2.1) are often taken together and used as indicators of areas' or communities' social well-being in the presence of tourism (e.g. Pizam, 1978; Perdue *et al.*, 1990; Ap and Crompton, 1998). Research into residents' perceptions of tourism impacts has been driven largely by three imperatives. Firstly, following on from the stage models of tourism development that evolved from the ethnographic case studies of the 1970s (Doxey, 1975; Smith 1978; Butler, 1980) and which suggested that residents' attitudes to tourism and tourists would deteriorate as levels of tourism increased, the social science research of the 1980s and 1990s contains attempts to test and measure this process. Secondly, researchers sought to evaluate empirically the negative effects of tourism reported in the ethnographic literature (see, for example, Pizam, 1978: 8) and to determine which characteristics of residents explained variations in perceptions of and attitudes to tourism and tourism development. Finally, by improving our understanding of community reactions to tourism, it would be possible, therefore, to: (a) determine areas' capacity for added tourism growth; and (b) take action to ameliorate or mitigate against (e.g. in future developments) residents' negative perceptions of tourism and tourists. In this way, tourists' experiences could be enhanced and tourism could be managed to operate at a level of *sustained equilibrium* within destination areas (Ap and Crompton, 1998: 123).

The findings from the various residents' perceptions studies have, however, revealed little consistency of results. Ap and Crompton (1998: 123), for example, note that contradictory findings have been reported for residents' perceptions of social impacts such as crime and vandalism, drug use and addiction, tourism's effect on the availability of recreational facilities and on family and social structures. Pearce *et al.* (1996: 19) also note that 'few consistent patterns or relationships were uncovered' by the social science based host community surveys of the type described above. However, they do go on to acknowledge that, from an analysis of the key findings and conclusions of research exploring residents' perceptions of impacts, some relationships can be seen. These relationships include (i) greater awareness of tourism impacts in areas of increased tourism development and (ii) attitudes to tourism impacts varying with: economic dependency or personal benefits gained from tourism; personal influence on tourism decision making; knowledge of tourism; and self-image and group identity.

In their summary of research into residents' reactions to tourism, Pearce *et al.* (1996: 27, 29) describe both the ethnographic tradition and the prevailing social science approach and identify three 'major problems in this area of research' (p.27). In the following section these problems are examined with specific reference to the Scottish research context upon which this paper is based and the author describes a novel conceptual approach, based upon findings from a review of the impact and wider literature which seeks to satisfy (at least some of) the concerns raised by Pearce *et al.* (1996) and others.

New Research Approaches

Shortcomings of the Existing Research

The first of the problems identified by Pearce *et al.* (1996: 27-8) relates to 'definitional and measurement problems with the concepts of tourists, tourism and community' (p. 27). They are concerned that the majority of studies have targeted differences in hosts while few have examined the effects of differing types of tourists or tourism. Indeed, this appears to be the case, despite the fact that as early as 1974, Butler suggested a framework for investigating tourism-induced social change which took account of the variable characteristics of tourists and tourism (see Table 2.1). The research under development by this author will address many of Butler's tourism/tourist characteristics through the choice of diverse case study areas. Additionally, recognition of Butler's (1974) tourism and tourist characteristics is implicit in the rationale for the work, which addresses the need for contextual sensitivity in dealing with tourism impacts and development issues in rural Scotland where, unlike many of the destination areas studied previously (and particularly those which formed the focus of the ethnographic studies of the 1970s), many visitors originate from within Scotland and the rest of the UK (typically only around 8 per cent of visitors to Scotland arrive from beyond the United Kingdom), and the development of tourism has taken place, not rapidly, but over an extended period of time.

Pearce *et al.* (1996: 27) go on to note a lack in consistency in how the term 'community' has been employed by tourism impact researchers and refer readers to Burr (1991) who found that impact researchers have often employed a place-based definition of community. While the usefulness of such spatially-based definitions of community is contestable (see, for example, Hillery, 1955; Stacey, 1969; Bell and Newby, 1971, 1976; and Harper, 1989), a place-based approach to establishing initial community membership appears reasonable in the context of this research. Such an approach is certainly practical as regards the design of a sampling framework for data collection. Additionally, in the context of this research – which seeks to identify community sub-groups based on residents' responses to the survey instrument – the benefits of a place-based approach to initially defining the community can be seen.

The second and third problems identified by Pearce *et al.* (1996: 27-8) are closely interrelated. The second of their problems relates to the way in which individual items for the measurement of residents' perceived impacts of tourism are selected: specifically, the issue lies in whether these are derived from residents' own perceptions of tourism impacts or from previous research undertaken elsewhere. Only in the former case, argue Pearce *et al.* (1996), can researchers be confident that the research results will reflect reality satisfactorily. That is, where residents can respond to interrogation based around impacts which have been generated during initial exploratory research undertaken in their community, rather than imposing upon them a set of *a priori* derived items to respond to. This distinction between the use of an *a priori* classification (the *etic*/positivistic approach) and the more exploratory (*emic*/interpretive) approach advocated by Pearce *et al.* (1996: 4) is reflected to some extent in the recently published debate published in the *Journal of Travel Research* (2001: 315-18) between Ap and Crompton – who posit an approach to tourism impact research which is strongly based on existing theoretical frameworks – and Lankford, who feels that research should be contextually sensitive and take into account '...variations in the level and content of development to reflect local concerns' (p. 316).

Although Lankford's earlier work describing the development of a tourism impact attitude scale (TIAS) (Lankford and Howard, 1994) did utilise an *a priori* set of items, those authors' interpretation of the factor structure derived from the data analysis (relating to community sub-groups) is criticised by Ap and Crompton (1998, p. 123) for not being consistent with the domains described by previous research. Lankford responds to this criticism by noting that similar findings had been reported elsewhere (2001: 315) and suggests that in some cases it may be more valid to seek new interpretations from the data rather than attempting to forcefully make the data analysis fit with existing theory.

The purpose of this discussion is not to take issue with either side in the debate described above, but to use it to help illustrate the dilemma faced by academics working in this, and other areas, of tourism research. Both Ap and Crompton, and Lankford have valid concerns. Dann *et al.* (1988), for example, found the development of tourism research in general to be limited by its atheoretical nature, whilst the potential limitations of restricting responses within tightly pre-defined frameworks have been illustrated by Schultz, who cautions that social reality should not be '...replaced by a fictional non-existing world constructed by the scientific observer' (1964: 8, cited in Ryan, 2000: 122). Concern relating to the atheoretical nature of tourism research is identified by Pearce *et al.* (1996: 27) as the third problem inherent in tourism impact research. Taking these second and third problems together, however, researchers may find themselves placed in a quandary: on the one hand Pearce *et al.* (1996) identify the need for contextual, or *emic*, sensitivity, while on the other hand they acknowledge the need for a less atheoretical approach to tourism research. Pearce *et al.* (1996: 4) describe *emic* approaches as 'drawing upon actors' interpretations', and contrast these with *etic* studies '...where the researchers generate their own constructs to describe the observed behaviour or cultural pattern'.

Viewing this research dilemma within the historical context of tourism impact research outlined above, a pattern can be described: many of the ethnographic studies of the 1970s employed, by their exploratory, qualitative nature, an *emic* approach to data collection, and from that work, stage model theories were generated to describe residents' responses to tourism development. Classically, these theories were then tested by researchers, working within what Jafari (1990) has described as the *adaptancy* and *knowledge-based* platforms, employing deductive reasoning and more quantitative social science research techniques to investigate empirically residents' perceptions and attitudes to tourism, tourism development and tourism impacts. However, despite such research being undertaken for over two decades, few consistent patterns or relationships have been uncovered (Pearce *et al.*, 1996: 25). Subsequently, in this area, as in other areas of tourism research, there have been calls in the literature for more exploratory, qualitative research to be undertaken – research which is more capable of explaining the complex social dimensions of the tourism phenomenon (e.g. Walle, 1997; Bowen, 2001; Dann and Phillips, 2001) or which can be used to generate new theoretical insights (e.g. Connell and Lowe, 1997).

Pearce *et al.*'s (1996) text can be viewed as bridging the *emic/etic* gap to some extent: their text proposes a *social representations* approach which they utilise both to theoretically ground their discussion on tourism impact research and as a framework for empirical analysis. In doing this, the social representations approach appears to enable researchers to adopt a theoretically explorative research approach while working within existing conceptual frameworks – as required by those in favour of more deductive research approaches. Such a 'mixed methodology' approach is not inconsistent with the way in which much research is conducted. Ryan (1995: 20-21), for example, describes the *functional approach* to research which embodies elements of both inductive and deductive reasoning; Veal

(1992: 30) also notes this approach. This paper now proceeds to outline an approach for studying the social impacts of tourism development in rural Scotland. This approach draws upon Pearce *et al.*'s (1996) description of social representations and the findings from a review of the tourism impact and wider tourism research literature. Within the context of the research dilemma described above, the strength of this new approach is that it offers the opportunity to satisfy the requirement for conducting research which contains an *emic*, exploratory element, within a wider conceptual framework which will allow existing theory to be strengthened accordingly. The element of *emic* sensitivity is required owing to the new and distinctive geographical context in which the research is being undertaken, while operating within existing theoretical frameworks addresses the need for tourism research to be less atheoretical in its nature.

Solutions and New Approaches

Pearce *et al.* (1996: 2) describe social representations theory as being:

> ...concerned with everyday knowledge and how people use this knowledge and common sense to understand the world in which they live and to guide their actions and decisions.

They aim to apply this theory to the study of the ways in which communities respond to tourism, in this way generating new theoretical insights into the relationship between tourism and communities; additionally, through secondary analyses of survey data relating to tourism impacts, planning and development, they aim to identify, within communities, clusters or networks of images which *can be viewed as* social representations, and which serve as a means of data reduction aiding, therefore, our understanding of tourism-community relationships. Pearce *et al.* (1996: 60) also write that:

> ...in order to understand how people react to tourism it will be valuable to understand the sources contributing to social representations.

They examine three such sources contributing to social representations: the printed and electronic media; social interaction; and individuals' direct experiences (pp. 83-91). It is suggested here, however, that this is too simplistic an analysis and one which does not take sufficient account of either the wider social science social representations models presented elsewhere in their text (pp. 31-57) or indeed the 'systems of benefits, values, attitudes and explanations which individuals and groups hold about tourism' which they refer to (p. 60).

Pearce *et al.* (1996: 3) citing the work of Moscovici (1981) write that social scientists must understand the world of everyday knowledge and common sense if they are to gain an understanding of social change. Building upon this sentiment, findings from previous research and a review of the tourism impact and wider tourism research literature, this paper describes a framework for research which seeks not only to clarify the social impacts of tourism in rural Scotland, but also how peoples' responses to tourism are influenced by wider social reality. Specifically, the focus of this proposed research will be on investigating individuals' *everyday-life aspirations* as potential determinants of how residents respond to tourism and form their attitudes to it. Everyday-life aspirations are viewed here as being: (i) a specific aspect of social representations which may act as determinants of residents' attitudes to tourism impacts and development; and (ii) informed by, and at the same time informing, social representations of tourism. This focus will, it is hoped, generate three specific outcomes: firstly it can be employed to explore the underlying factors responsible for variations in residents' perceptions of and attitudes towards tourism; secondly, it will enable research to be undertaken within existing theoretical frameworks while maintaining an exploratory, or *emic* approach to the work in order to strengthen existing theory with new insights; and, thirdly, it will help to develop and theoretically underpin the social representations approach to community tourism relationships by providing data relating to the sources contributing to social relations. Additionally, in this way, the research will provide an important contribution to the literature concerned with the planning and management of tourism development specifically in the context of rural areas of Scotland.

The Everyday Life Approach

Relativity   Anecdotal evidence from previous research undertaken by this author suggested that residents' attitudes to tourism may be influenced by the status of tourism relative to other salient factors in individuals' and communities' lives. Specifically, during fieldwork undertaken in 1997 that investigated residents' attitudes towards tourism employment in rural Scotland, it was noted that the greatest number of positive responses relating to tourism employment and to tourism in general were recorded on the Clyde Coast island of Bute which, of all the case study areas surveyed, had suffered the most dramatic downturn in visitor numbers during the two decades previous to the research being undertaken. The tourism-related sector had been responsible for economic and social vibrance on the island and the demise of this sector (along with the decline experienced in the agricultural sector) had led to Bute to suffer the severest and most consistent decline throughout the Highlands and Islands since the 1960s (BPSG, 1994: 1, cited in Boyne *et al.*, 2000: 106). In a similar vein, Perdue *et al.* (1990) reporting on their Colorado research suggested that support for tourism was higher in communities pessimistic about their economic future and referred to this as the 'doomsday phenomenon' (cited in Getz, 1994: 248).

Considering Doxey's (1975) Irridex or irritation index (which posits an inverse relationship between levels of hosts' irritation with tourists and the growth of tourist numbers over time), it may be that as a destination goes into decline and tourist numbers decrease, residents' attitudes to tourists and tourism become (once again) more positive. Doxey's relationship between tourist numbers and residents' attitudes may not, therefore, be uni-linear in its nature. Indeed Young *et al.* (1999)

have noted a similar variation in residents' attitudes occurring on a seasonal basis. Perhaps, therefore, it may be useful to extend Doxey's model to include a 'nostalgia' stage (see Figure 2.3). Although this modification requires us to rename Doxey's *Final Level* as the *Penultimate Level*, it should be noted that this theoretical model has no end-point – but rather, hosts' attitudes are subject to ongoing change dependent on fluctuating tourist numbers.

Other areas in which residents may form their reactions to tourism based on its status relative to other factors may include: level of tourism development (already well conceptualised and tested – but relationship not found to be consistent); the desirability of tourism employment, not only at individual and societal levels, but also relative to other employment opportunities; the rate of growth of tourism; and, the length of time that tourism has been in existence in the destination area.

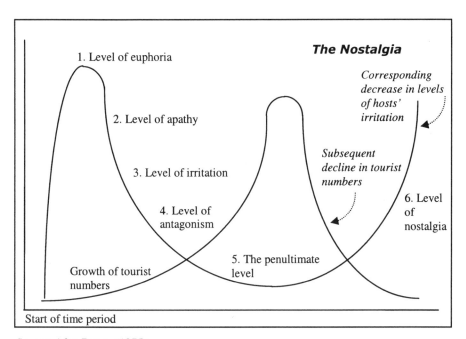

Source: After Doxey (1975)

Figure 2.3   The modified/extended Doxey's Irridex model

Individuals' aspirations Within the free-market liberalism paradigm of economic growth and development, it is often held that most individuals favour economic growth and that tourism-related developments promoting the generation of wealth will, therefore, meet with approval from areas' residents.   In his discussion on development theory, Sharpley (2000: 4) draws our attention to Goulet's (1968) work who writes that '...development can be properly assessed only in terms of the total needs, values, and *standards of the good life and the good society* perceived by the very societies undergoing change' (emphasis added). It follows from this logic, then, that responses to development will also be influenced by individuals' 'needs, values, and standards of the good life'. From their comparative analysis of residents' attitudes to tourism development in the UK and New Zealand, Ryan *et al.* (1998) found that residents in more mature destination areas are more likely to express concern about the nature of tourism and its development. Additionally, from this work, Ryan *et al.* confirmed the findings of previous research where socio-demographic discriminators failed to explain variations in residents' support or opposition to tourism development. As a framework for future research, they propose a value-orientated approach to understanding the factors which act as determinants of residents' reactions to tourism. They suggest that in the early stages of tourism development, residents' core values are not used to evaluate tourism, as tourism appears to contribute to economic well-being with few, if any, associated costs being evident. As development intensifies and tourism numbers increase, and impacts become more apparent, then there is a need for residents to become more discerning regarding tourism and they begin to apply their more deeply-seated core values resulting in more circumspect reactions.

In the examples above, 'standards of the good life' and 'core values' are shown as variables which can influence the way in which people perceive development processes. These variables can be viewed as everyday life factors which contribute to and are informed by social representations. The notion that 'everyday life' aspirations can underlie residents' reactions is further backed up by Madrigal (1995, p. 100) where he notes that different nested communities will share one or more characteristics in common; specifically, he cites *needs* and *wants* as two examples of shared characteristics. Elsewhere, Snepenger and Johnson (1991) write that political self-identification has been shown to be a key lifestyle indicator (p. 512) and found that it could be a significant correlate in how individuals perceive tourism. The above can be viewed here as examples of everyday-life aspirations affecting residents' attitudes towards tourism development. Two further ways in which a focus on everyday life can contribute towards furthering our understanding of communities' reactions to tourism are described below.

The role of everyday life in social science analysis Pearce *et al.* (1996) describing the theory of social representations write that it is 'concerned with understanding everyday knowledge' (p. 2) and refer to Moscovici (1981) noting that 'social scientists must understand *the world of everyday knowledge* and common sense if they are to gain an understanding of social change' (p. 3; emphasis added). They go on to reinforce this describing how an *emic* approach can assist the search for clusters of images (which would be indicative of the existence of social representations) by helping to 'explore the range of attitudes and breadth of residents' opinions' (p. 61). Additionally, at this point in their text, they suggest that rather than overall attitude to tourism being created from perceptions of its impacts (the basis of much residents' attitudes research), it may be the case that the way in which impacts are perceived and felt is structured by peoples' overall image of tourism and tourists *and their associated beliefs*. Further evidence for the potential merit of undertaking residents' attitudes research utilising an 'everyday-life' theoretical framework is provided in the following analysis of Ryan's (2000) paper which describes a phenomenographic approach to understanding consumer satisfaction in the context of tourism.

Phenomenography (n.b. not phenomenology), argues Ryan, is highly appropriate for being applied to tourism studies as it stresses the relationship between the individual and the experience. Phenomenography is the study of how people perceive the world: it is an *emic* approach in that it is concerned with 'eliciting the individual's construction' and seeks 'an understanding of what constitutes the structures of conceptualising a phenomenon' (Ryan, 2000, p. 123). Ryan suggests that 'Subject and object exist not in a dichotomous relationship, but one which is reiterative' (p. 122). To illustrate this relationship he describes an individual on holiday in Majorca (p. 123). Turning Ryan's holiday in Majorca scenario upside down to focus on the resident receiving the tourist, however, generates the following proposition:

> ...for residents to have tourists arrive in their town is for them as hosts to experience tourism through those tourists and their interactions with them. The tourists are context specific (they are in the hosts' town), and the hosts' learning (perceiving impacts/coping with tourism) relates to the experience of having those tourists in their place. The experience, however, is also an experience of something called 'everyday life' – an experience of a structure of meaning *that is very place specific*. There is [*extrapolating from Ryan's reference to Husserl, 1970/1901*], a relationship between the objective (tourism in the hosts' town) and the subjective (residents' perceptions of tourism's impact): the two components (tourism/objective and impact perception/subjective) form a whole.
>
> Hence any analysis that separates tourism (e.g. a categorisation of tourism impacts) and loci of experience (hosts' wishes and aspirations for a high quality of everyday life) is incomplete in developing an understanding of, say, 'hosts' everyday life with tourism in their town'.

This proposition strengthens the rationale for seeking to establish everyday-life factors as discriminants for understanding differences in respondents' reactions to tourism. Building upon Ryan's (2000) phenomenographic approach to understanding consumer satisfaction in tourism, and the insights from the literature described above, it is hoped that by successfully employing the 'everyday-life' approach to seek underlying factors which explain variations in the ways in which residents respond to tourism's impacts, a precedent can be set for future empirical research in this area to build upon while also strengthening the underpinning theoretical frameworks.

Summary and Conclusions

The section above has described the theoretical context for the proposed everyday-life approach. It is argued here that peoples' everyday-life aspirations form a major source contributing to individuals' and communities' social representations. Everyday-life aspirations will both be informed by and inform social representations; in this respect they will act as determinants affecting residents' attitudes to tourism. Three specific aspects of everyday life and its potential interplay with how residents perceive and react to tourism development and its associated impacts are described. Firstly, residents will react to tourism relative to other industrial activity in their area and the wider economic well-being in their community or region. Secondly, individuals will react based on the values they hold: these values may relate to development in general, tourism development in particular, and/or the ways in which they conceive a high standard of life and may be related to political self-identification. Finally, drawing upon social representations theory, the importance of an understanding of everyday life in social science research is highlighted. This is exemplified further by adapting Ryan's (2000) phenomenographic approach to studying consumer satisfaction to focus on the study of resident reactions to tourism.

The paper also highlights the tension which is apparent in the tourism impact and residents' perceptions research literature where it relates to the epistemology of enquiry. Specifically, this tension is manifest in the calls on the one hand for research to be undertaken in a strongly deductive manner (e.g. theory testing within existing conceptual frameworks) while on the other hand, some academics argue that more exploratory, qualitative or emic research approaches are required to understand tourism's complexity and to maintain contextual sensitivity through the avoidance of overly prescriptive data collection methods. Within this context, this paper describes a novel conceptual approach to seeking commonalities in residents' responses which is embedded within the established wider theoretical framework relating to tourism impact research. Such an approach reflects the reiterative nature of the wider tourism research tradition (Ryan, 1995: 20) and acknowledges Walle's (1997: 535) 'plurality of equally valid research strategies'. Relating the above observations to Jafari's (1990) platform model of tourism research, it is suggested here that the tourism impact/residents' perceptions research (and tourism studies in general) may be witnessing the emergence of a new, fifth, research platform; this can be regarded as a post-positivistic[1] platform, where research is undertaken from a variety of epistemological perspectives and in which there is an ongoing natural selection of knowledge. Of course, if we consider the fragmentary nature of both historical and contemporary tourism research (impact-specific and in general), we may conclude that the post-positivistic platform has been in existence and operation for some time.

Note

1 The author refers to post-positivism as an epistemic perspective which, while rejecting the absolute authority of the hypothetico-deductive approach, recognises the validity of positivistic modes of enquiry where these are appropriate.

References

Ap, J. (1990), 'Residents' Perceptions Research on the Social Impacts of Tourism', *Annals of Tourism Research*, vol. 17, pp. 610-16.

Ap, J. (1992), 'Residents' Perceptions on Tourism's Impacts', *Annals of Tourism Research*, vol. 19, pp. 665-90.

Ap, J. and Crompton, C. (1998), 'Developing and Testing a Tourism Impact Scale', *Journal of Travel Research*, vol. 37, pp. 120-30.

Ap, J. and Crompton, C. (2001), 'Response to Lankford', *Journal of Travel Research*, vol. 39, pp. 317-18.

Belisle, F.J. and Hoy, D.R. (1980), 'The Perceived Impact of Tourism by Residents: A Case Study in Santa Marta, Columbia', *Annals of Tourism Research*, vol. 7, pp. 83-101.

Bell, C. and Newby, H. (1971), *Community Studies: An Introduction to the Sociology of the Local Community*, Allen and Unwin, London.

Bell, C. and Newby, H. (1976), 'Community, Communion, Class and Community Action', in D.T. Herbert and R.J. Johnston (eds), *Social Areas in Cities*, John Wiley & Sons, Chichester, pp. 189-207.

Bowen, D. (2001), 'Research on Tourist Satisfaction and Dissatisfaction: Overcoming the Limitations of a Positivist and Quantitative Approach', *Journal of Vacation Marketing*, vol. 7 (1), pp. 31-40.

Boyne, S., Hall, D. and Gallagher, C. (2000), 'The Fall and Rise of Peripherality: Tourism and Restructuring on Bute', in F. Brown and D. Hall (eds), *Tourism in Peripheral Areas*, Channel View Publications, Clevedon, pp. 101-13.

BPSG (Bute Partnership Steering Group) (1994), *Bute Action Plan: A Development Strategy for the Isle of Bute*, BPSG, Rothesay.

Brougham, J.E. and Butler, R.W. (1981), 'A Segmentation Analysis of Resident Attitudes to the Social Impact of Tourism', *Annals of Tourism Research*, vol. 8, pp. 569-90.

Brown, F. (1998), *Tourism Reassessed: Blight or Blessing?* Butterworth-Heinemann, Oxford.

Brown, G. and Giles, R. (1994), 'Coping With Tourism: An Examination Of Resident Responses To The Social Impact Of Tourism', in A.V. Seaton, C.L. Jenkins, P.U.C. Dieke, M.M. Bennet, L.R. MacLellan and R. Smith (eds), *Tourism: The State of The Art*, John Wiley & Sons, Chichester, pp. 755-64.

Brunt, P. and Courtney, P. (1999), 'Host Perceptions of Sociocultural Impacts', *Annals of Tourism Research*, vol. 26, pp. 493-515.

Burr, S.W. (1991), 'Review and Evaluation of the Theoretical Approaches to Community as Employed in Travel and Tourism Impact Research on Rural Community Organisation and Change', in A.J. Veal, P. Jonson and G. Cushman (eds), *Leisure and Tourism: Social and Environmental Changes*, Papers from the World Leisure and Recreation Association Congress, Sydney, Australia.

Butler, R.W. (1974), 'The Social Implications of Tourism Developments', *Annals of Tourism Research*, vol. 2, pp. 100-12.

Butler, R. (1980), 'The Concept of a Tourism Area Life Cycle of Evolution', *Canadian Geographer*, vol. 24, pp. 5-12.

Cohen, E. (1978), 'The Impact of Tourism on the Physical Environment', *Annals of Tourism Research*, vol. 5, pp. 215-37.

Connell, J. and Lowe, A. (1997), 'Generating Grounded Theory from Qualitative Data: The Application of Inductive Methods in Tourism and Hospitality Management Research', *Progress in Tourism and Hospitality Research*, vol. 3, pp. 165-73.

Dann, G. (1996), *The Language of Tourism: A Sociolinguistic Perspective*, CAB International, Wallingford.

Dann, G., Nash, N. and Pearce, P.L. (1988), 'Methodology in Tourism Research', *Annals of Tourism Research*, vol. 15, pp. 1-28.

Dann, G. and Phillips, J. (2001), 'Qualitative Tourism Research in the Late Twentieth Century and Beyond', in B. Faulkner, G. Moscardo and E. Laws (eds), *Tourism in the Twenty-First Century: Reflections on Experience*, Continuum Publishers, London, pp. 247-65.

Davis, D., Allen, J. and Cosenza, M. (1988), 'Segmenting Local Residents by their Attitudes, Interests and Opinions Towards Tourism', *Journal of Travel Research*, vol. 27, pp. 2-8.

de Kadt, E. (ed.) (1979), *Tourism: Passport to Development? Perspectives on the Social and Cultural Effects of Tourism in Developing Countries*, Oxford University Press, New York.

Dogan, H.Z. (1989), 'Forms of Adjustment: Sociocultural Impacts of Tourism', *Annals of Tourism Research*, vol. 16, pp. 216-36.

Doxey, G. (1975), *A Causation Theory of Visitor-Resident Irritants: Methodology and Research Inferences*, The Impact of Tourism. Proceedings of the 6th Annual Conference of the Travel Research Association, San Diego, CA, Travel and Tourism Research Association, pp. 195-8.

Finney, B.R. and Watson, K.A. (eds) (1977), *A New Kind of Sugar: Tourism in the Pacific*, Santa Cruz: Centre for South Pacific Studies.

Fredline, E. and Faulkner, B. (2000), 'Host Community Reactions: A Cluster Analysis', *Annals of Tourism Research*, vol. 27, pp. 763-84.

Getz, D. (1994), 'Residents' Attitudes Towards Tourism: A Longitudinal Study in the Spey Valley, Scotland', *Tourism Management*, vol. 15 (4), pp. 247-58.

Goulet, D. (1968), 'On the Goals of Development', *Cross Currents*, vol. 18, pp. 387-405.

Harper, S. (1989), 'The British Rural Community: An Overview of Perspectives', *Journal of Rural Studies*, vol. 5, pp. 161-84.

Hillery, G.A. (1955), 'Definitions of Community: Areas of Agreement', *Rural Sociology*, vol. 20, pp. 111-23.

Husserl, E. (1970/1901), *Logical Investigations: Vol. 2*, Routledge and Kegan Paul, London (translated by J.N. Findlay from German, *Logische Untersuchungen*).

Jafari, J. (1990), 'Research and Scholarship: The Basis of Tourism Education', *Journal of Tourism Studies*, vol. 1, pp. 33-41.

Lankford, S.V. (2001), 'A Comment Concerning "Developing and Testing a Tourism Impact Scale"', *Journal of Travel Research*, vol. 39, pp. 315-16.

Lankford, S.V. and Howard, D.R. (1994), 'Developing a Tourism Impact Attitude Scale', *Annals of Tourism Research*, vol. 21, pp. 121-39.

Madrigal, R. (1995), 'Residents' Perceptions and the Role of Government', *Annals of Tourism Research*, vol. 22, pp. 86-102.

Mathieson, A. and Wall, G. (1982), *Tourism: Economic, Physical and Social Impacts*, Longman, Harlow.

Moscovici, S. (1981), 'On Social Representations', in J.P. Forgas (ed.), *Perspectives on Everyday Understanding*, Academic Press, London, pp. 181-209.

Pearce, P.L., Moscardo, G. and Ross, G.F. (1996), *Tourism Community Relationships*, Elsevier Science, Oxford.

Perdue, R.R., Long, P.T. and Allen, L. (1990), 'Resident Support for Tourism Development', *Annals of Tourism Research*, vol. 14, pp. 420-9.

Pizam, A. (1978), 'Tourism Impacts: The Social Costs to the Destination Community as Perceived by its Residents', *Journal of Travel Research*, vol. 16, pp. 8-12.

Prentice, R. (1993), 'Community-driven Tourism Planning and Residents' Preferences', *Tourism Management*, vol. 14, pp. 218-27.

RDC (Rural Development Commission) (1996), *The Impact of Tourism on Rural Settlements*, Rural Research Report No. 21, Rural Development Commission, Salisbury.

Runyan, D. and Wu, C.T. (1979), 'Assessing Tourism's More Complex Impacts', *Annals of Tourism Research*, vol. 6, pp. 448-63.

Ryan, C. (1995), *Researching Tourist Satisfaction: Issues, Concepts, Problems*, Routledge, London.

Ryan, C. (2000), 'Tourist Experiences, Phenomenographic Analysis, Post-positivism and Neural Network Software', *International Journal of Tourism Research*, vol. 2, pp. 119-31.

Ryan, C., Scotland, A. and Montgomery, D. (1998), 'Resident Attitudes to Tourism Development: A Comparative Study between the Rangitikei, New Zealand and Bakewell, United Kingdom', *Progress in Tourism and Hospitality Research*, vol. 4, pp. 115-30.

Schultz, A. (1964), *Studies in Social Theory*, Martinus Nijhoff, The Hague.

Sharpley, R. (2000), 'Tourism and Sustainable Development: Exploring the Theoretical Divide', *Journal of Sustainable Tourism*, vol. 8, pp. 1-19.

Sheldon, P.J. and Var, T. (1984), 'Residents' Attitudes to Tourism in North Wales', *Tourism Management*, vol. 5, pp. 40-47.

Smith, V. (ed.) (1978), *Hosts and Guests: The Anthropology of Tourism*, Blackwell, Oxford.

Snepenger, D.J. and Johnson, J.D. (1991), 'Political Self-identification and Perceptions on Tourism', *Annals of Tourism Research*, vol. 18, pp. 511-15.

Stacey, M. (1969), 'The Myth of Community Studies', *British Journal of Sociology*, vol. 20, pp. 134-7.

Turner, L. and Ash, J. (1975), *The Golden Hordes: International Tourism and the Pleasure Periphery*, Constable, London.

UNESCO (United Nations Educational, Scientific and Cultural Organisation) (1976), 'The Effect of Tourism on Socio-cultural Values', *Annals of Tourism Research*, vol. 4, pp. 74-105.

Veal, A.J. (1992), *Research Methods for Leisure and Tourism: A Practical Guide*, Pitman, London.

Walle, A.H. (1997), 'Quantitative Versus Qualitative Tourism Research', *Annals of Tourism Research*, vol. 24, pp. 524-36.

Wheat, S. (1994), 'Is There Really an "Alternative" Tourism?', *Tourism in Focus*, vol. 13, pp. 2-3.

Wheeller, B. (1991), 'Tourism's Troubled Times: Responsible Tourism is Not the Answer', *Tourism Management*, vol. 12, pp. 91-6.

Williams, S. (1998), *Tourism Geography*, Taylor and Francis Books Ltd., London.

Young, G. (1973), *Tourism: Blessing or Blight?* Penguin, Harmondsworth.

Young, T., Thyne, M. and Lawson, R. (1999) 'Comparative Study of Tourism Perceptions', *Annals of Tourism Research*, vol. 26, pp. 442-5.

Chapter 3

Rural Tourism and Sustainability – A Critique

Richard Sharpley

Introduction

Over the last fifteen years, the concept of sustainable tourism development has become almost universally accepted as a desirable and politically appropriate approach to, and goal of, tourism development. It has achieved 'virtual global endorsement as the new [tourism] industry paradigm' (Godfrey, 1996: 60) and, from the local to international level, innumerable sectoral and destinational organisations within the tourism industry have produced sustainable tourism development plans and policies. Tourism academics have, likewise, devoted much attention to the subject of sustainable tourism from both an operational and definitional point of view. Indeed, it has been observed that the 1990s are particularly memorable for the volume of literature related to the topic (Yeoman, 2000) whilst 'defining sustainable development in the context of tourism has become something of a cottage industry' (Garrod and Fyall, 1998).

Such widespread concern with the concept is not surprising. Not only has the emergence of environmentalism as a dominant global political and social movement since the late 1960s added a new environmental dimension to most economic, political and social activities (Yearley, 1992), but also understanding of and approaches to the processes and goals of development itself have undergone a significant transformation (Sharpley, 2002a). As a result, sustainable development has emerged as the dominant development paradigm, driving both global policy-making and strategy in general (WCED, 1987) as well as informing more specific sectoral policies and activities including, of course, those related to tourism (WTO/WTTC, 1996). This has been particularly evident in the rural context. As tourism has become increasingly viewed as an effective means of addressing the socio-economic challenges facing peripheral rural regions (Cavaco, 1995; Hoggart *et al.*, 1995; Oppermann, 1996), sustainability – both optimising the developmental benefits of tourism and satisfying the needs of tourists within strict environmental parameters – has become a dominant principle and objective. Indeed, rural tourism and sustainable tourism have become virtually synonymous, with sustainability an integral element of many rural tourism development policies. In the UK, for example, the former Countryside Agency published a set of principles for sustainable rural tourism (Countryside Commission, 1995) whilst the most recent policy document (ETC/CA, 2001) also focuses primarily upon sustainable tourism development.

However, despite the widespread support for the principles and objectives of sustainable rural tourism development, it remains a contested concept. Not only is it variously interpreted (Hunter, 1995), but also its validity as means and objective of tourism development is being increasingly questioned in many quarters. These arguments are well rehearsed in the literature and are, in general, beyond the scope of this chapter (see Wall, 1997; Roberts and Hall, 2001; Sharpley, 2000; 2002b). Nevertheless, two specific factors that have served to raise significant questions over the viability of sustainable (rural) tourism development deserve attention. Firstly, there has been increasing recognition that, although the adoption of its principles may achieve certain economic, marketing and public relations objectives for individual tourism businesses (Butler, 1998) and that, under certain circumstances, it represents the most appropriate form of tourism development, it does not represent a universally applicable framework for developing tourism. In other words, the principles of sustainable tourism are increasingly viewed as a development 'blueprint' that is unable to accommodate the almost infinite diversity of tourism development contexts (Southgate and Sharpley, 2002).

Secondly, with specific reference to rural tourism in Britain, the events of 2001 served to highlight the fragility of tourism as a sector of rural economies. That is, the outbreak of foot and mouth disease (FMD) in late February that year and the subsequent, virtually instantaneous collapse of the rural tourism industry, both in the areas worst affected by the disease more generally around the countryside, were evidence of not only the interdependence between tourism and the wider rural economy and society but also the inherent lack of sustainability of rural tourism. In short, an industry which is widely promoted as a means of developing or regenerating rural economies and societies and which is intimately associated with the notion of sustainable development was rapidly brought to its knees by events beyond its control.

These two factors are related inasmuch as the FMD crisis verified a number of criticisms levelled at the sustainable rural tourism thesis. More specifically, it highlighted a fundamental problem that challenges the concept of sustainable tourism in general, namely, the issue of control or governance. That is, an implicit requirement for the achievement of sustainable tourism development is a local, community approach which seeks to optimise the benefits of tourism to all stakeholders but according to local needs. However, recent experience in Britain has demonstrated that not only is there little control at the local level but also that, at the national level, rural tourism policy is dictated by, and subordinated to, broader rural development objectives.

The purpose of this chapter, therefore, is to explore the implications of the FMD crisis for rural tourism development, focusing, in particular, on the governance of the tourism sector. In so doing, it not only adds substance to a number of criticisms directed at the concept of sustainable rural tourism but, more positively, it seeks to identify lessons for the future role of tourism in sustainable rural development. Specifically, it argues that, at a national level, rural tourism policy in particular must be more closely aligned with rural development policy in general. The first task, then, is to briefly consider the notion of governance in the rural context and how it relates to sustainable tourism development.

Rural Governance and Sustainable Tourism

As suggested above, an implicit characteristic of sustainable tourism development is localisation.[1] That is, sustainable tourism development should, ideally, focus on satisfying the development needs of local communities through, for example, promoting local product supply chains, encouraging local crafts and industries, optimising the retention of tourism earnings within the destinations and ensuring that development is within local environmental and social capacities. The achievement of such localisation is, in turn, dependent upon local control of tourism development. Indeed, a fundamental objective of sustainable development is the satisfaction of basic needs and the encouragement of self-reliance based upon a grassroots, endogenous developmental process (Streeten, 1977; Galtung, 1986). The viability of such an approach must, however, be considered within the framework of broader political, economic and social structures. To put it another way, the nature of the wider political economy should be conducive to the local planning and control of all economic sectors and activities, though co-ordinated at the regional and national level in order to meet broader developmental objectives.

Importantly, in recent years this broader political economy has, arguably, become increasingly complex in the specific context of the countryside (Marsden and Murdoch, 1998). In particular, a preoccupation with agricultural interests and concerns, at both the local and national governmental level, has been superseded by a more diverse, fragmented approach to the governing of rural areas. As Marsden and Murdoch (1998) suggest, rural government has now 'splintered into a multitude of political processes', reflecting the declining role of the agricultural sector in the contemporary countryside combined with the increasingly diverse range of demands made, including tourism, upon rural spaces.

At the same time, however, the wider political context within which rural government operates has itself undergone a transformation, manifested in a shift from traditional 'government' to 'governance' (Goodwin, 1998). Whereas the former relates to the more formal institutional structure and location of authoritative decision making in the modern state, the latter is a broader concept, referring to the 'distribution of power both internal and external to the state' (Stoker, 1997: 10). In particular, governance is concerned with the plurality of governmental and non-governmental institutions and organisations working together to address particular social and economic issues and challenges and is, perhaps, most clearly evident in the changing emphasis of the public intervention *versus* free market debate. Thus, the neo-liberal, free market approach of the 1980s and 1990s (as the antithesis to old style public intervention) has been replaced with the 'third way' – an attempt to effectively integrate state-sponsored redistribution with a market led economy. The concept of governance is summarised by Goodwin (1998) who suggests that it:

- refers to a complex set of institutions and actors that are drawn from but also beyond government;
- identifies the blurring of boundaries and responsibilities for tackling social and economic issues;
- identifies the power dependence in the relationship between institutions involved in collective action;
- is about autonomous self-governing networks of actors; and
- recognises the capacity to get things done which does not rest on the power of government to command or use its authority.

The extent to which governance, as opposed to government, is an effective decision-making and policy-implementation process remains to be seen – certainly, as argued here, the experience of the FMD crisis would suggest that, ultimately, central government dominance is inevitable. Nevertheless, in the context of this chapter it provides a useful theoretical foundation for analysing the complex structure and inter-relationships between the enormous range of public, private and voluntary organisations involved in governing the countryside in general, and for considering the viability of sustainable rural tourism in particular.

There is no doubt that the governance concept is an effective framework for mapping the complex institutional structure of control and responsibility in rural areas in terms of both overall governance and also at the level of specific sectors and processes. Not only does the countryside play host to an increasingly diverse array of stakeholders, all of whom have valid claims on the rural resource, but also – as the FMD crisis emphasised – an interdependence exists between all of these stakeholders. Moreover, at the sectoral level, a governance approach (frequently manifested in 'partnerships') has become increasingly evident. In the tourism context, for example, Tourism Development Partnerships, representing an amalgam of public, private and local community interests, have been an integral element of local tourism planning in both urban and rural areas since the late 1980s. Similarly, schemes such as the LEADER EU rural development programme are evidence of a governance approach to rural tourism development at a regional level.

Importantly, however, it is necessary to go beyond simply describing the emergence of new structures and mechanisms of rural governance. To understand the ways in which policies are determined or courses of action are decided upon (or, indeed, to appreciate or identify those forces and influences within a particular sphere of governance that facilitate or militate against the achievement of specific objectives), it is essential to question the roles, interests and rationales of relevant agencies and organisations. In other words, all actors within a particular governance context will embrace differing ideologies and purposes with respect to the development or exploitation of the rural resource base. Therefore, as Goodwin (1998) proposes, 'the task is to interrogate the reasons both for their emergence and for their outputs. What exactly do they do, and why?' As the next section considers, such an enquiry may reveal factors that, in the context of this paper, support the position that 'true' sustainable rural tourism development is not a viable objective.

The Governance of Rural Tourism: Competing Ideologies

The modern countryside has become an arena in which a multitude of tensions and competing demands are played out, frequently reflecting wider social and economic differences and conflicts. Issues such as access (Shoard, 1999), equal opportunities for participation in rural recreation (Harrison, 1991), the gentrification of rural areas and competing perspectives on development, land use and, more generally, what the countryside 'is', are all high on the rural political agenda whilst, at a more micro level, appropriate recreational use of the countryside remains the subject of intense debate. However, despite the enormous variety of organisations and agencies involved in the planning, management and exploitation of the countryside, it is possible, according to Stone (2000), to categorise their activities, purpose and direction under the heading of one of two underlying ideological perspectives.

On the one hand, a dominant ideology followed by many organisations in the public sector, including the Countryside Agency, as well as landowners and, arguably, the middle classes, is the so-called 'countryside aesthetic' (Harrison, 1991), an ideology driven by the desire to maintain a picturesque rural idyll, a green and pleasant land (Newby, 1985) which, in the tourism context, supports 'appropriate' quiet or solitary recreational pursuits. To this group can be added all the organisations concerned with the protection or preservation of the countryside, its environment, wildlife and culture, to collectively form what can be termed the conservation ideology.

On the other hand, innumerable organisations and businesses are, of course, concerned with the commercial exploitation of the rural resource. They adopt a rational, as opposed to idyllic, approach to the countryside, viewing it as a resource from which a profit can be earned. In the tourism context, this group of actors includes not only the rural tourism 'industry' – that is, all those businesses that directly and indirectly depend upon the rural resource – but also certain public sector agencies, such as tourist boards, whose prime objective is the profitable development of tourism. Together, these organisations and businesses are influenced by a dominant 'commercial' ideology (Stone, 2000).

Inevitably, distinctions between these two competing ideologies are not always clear, particularly in the rural tourism context. For example, the success of any rural tourism enterprise depends upon the maintenance of a healthy and attractive rural environment, hence the commercial logic of striving to achieve environmentally sustainable resource exploitation (as distinct from sustainable development). Nevertheless, the concept of two competing ideologies provides a useful framework for developing a model of the governance of sustainable rural tourism, as shown in Figure 3.1. This proposes that the successful governance of rural tourism is dependent upon the planning and management of tourism, whether through a partnership of relevant stakeholders or through an institution invested with appropriate authority, which seeks to recognise, satisfy and balance the opposing needs of the conservation and commercial groups of actors within broader developmental objectives.

A number of points deserve mention:

- whilst such a balance between the competing needs and ideologies of stakeholders may in practice be possible, it is only likely to occur at the very local level (i.e. within the context of specific projects or developments involving relatively few participants). Such is the potential number and diversity of stakeholders at a broader context, whether local or regional, that consensus is unlikely to be achieved. This, perhaps, explains the fact that, in reality, sustainable rural tourism is manifested in small scale projects and, even then, may not satisfy all the requirements for 'true' sustainable development;

- although governance implies operations beyond the control of formal structures of government, it is, nevertheless, impossible to divorce the governance process from the influence of central government and national policy-making. For example, public sector agencies, though not directly controlled by government, remain dependent upon the public purse for funding and their activities are often constrained by national policy; in the case of tourism, the responsibilities of the national and regional tourist boards and the extent to which they can support tourism is significantly restrained by their defined roles and their levels of funding. At the same time, through its various policies with respect, for example, to employment and taxation, central government establishes the broader parameters within which any sector, including tourism, must operate;

- related to both of the above points, and given the complexity of political structures in the rural context where an inter-dependency exists between the different demands upon the rural resource, the potential for achieving overall sustainable rural development may be impossible to achieve. In other words, policies exist at the national level for the different rural activities and sectors, such as farming, forestry and tourism, and it is inevitable that policies related to one sector will inevitably impact on another. As this chapter now argues, this was particularly evident in the FMD crisis, during which the needs of rural tourism were clearly subordinated to national agricultural policy.

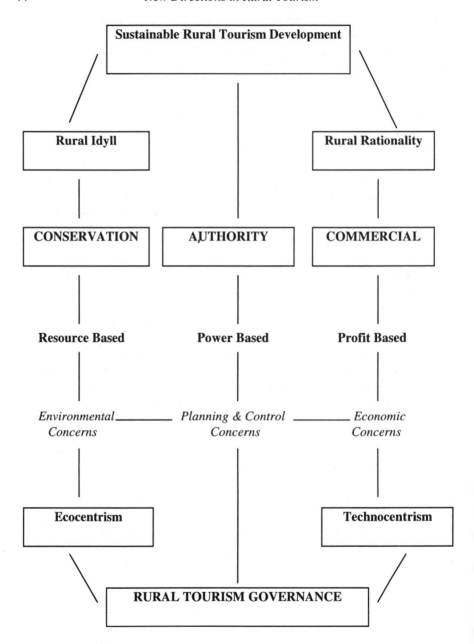

Source: Stone (2000)

Figure 3.1 A model of sustainable rural tourism governance

Tourism and the Foot and Mouth Disease Crisis: An Overview

A full consideration of the 2001 FMD crisis and its consequences for rural tourism is beyond the scope of this paper (see, for example, Sharpley and Craven, 2001; Blake *et al.*, 2002). Nevertheless, a brief review of the principal impacts of the crisis and the government's policy responses, both in attempting to control the disease and subsequently to alleviate the problem facing the tourism sector, reveals significant implications for the governance of rural tourism and for sustainable rural development.

The first case of FMD was confirmed at an abattoir in Essex on 21 February 2001, signalling not only the first major outbreak of the disease in Britain since 1967, but also the onset of one of the most serious economic and social crises to face rural communities in recent years. Indeed, despite government assurances as early as 12 March that the outbreak was under control, it was not until the end of that month that the daily number of new cases peaked and, even by mid-June, a small number of new cases were being confirmed each day. By that stage, 3,290,000 animals had been slaughtered and 1,740 cases of the disease had been confirmed although, as a direct result of the government's policy, announced on 24 March, to cull all livestock within a three kilometre radius of infected farms, a total of 8,100 farms and other premises had been affected.

There can be no doubting the severe and far-reaching consequences of the outbreak for an agricultural sector already suffering significant economic sad social hardship. However, the costs of the crisis to the wider economy were far greater than those borne by the farming sector. In particular, the tourism industry endured a huge downturn in business, with estimates of losses ranging from £140mn to £500mn a week (ETC, 2001a; Brown, 2001) and, for England alone, a potential overall loss of over £5bn in 2001 (ETC, 2001b).

These losses resulted, to a great extent, not from the spread of the disease itself, but the policies designed to control it. Initially – given the nature of FMD as a highly contagious disease which can be transmitted through contact with infected animals, animal products, contaminated people or equipment, or on the wind – the first measures taken were designed to contain the spread of the disease (CMSC, 2001). An immediate ban was placed on all movements of animals, including a ban on the export of live animals and by 23 February, the government was advising the general public to avoid contact with livestock farms, walking in areas where there were livestock, and any unnecessary visits to farmland.

At the same time, however, a number of more specific controls were also imposed to minimise the risk of the general public (i.e. tourists) spreading the disease. Not only were the majority of footpaths and other rights of way closed across the country, but also many tourist attractions and destinations, including forests, some national parks and even parks in south-west London, were also closed. In effect, the government's initial response to the outbreak, in addition to specific controls on the farming sector, was to place the entire British countryside under quarantine.

To an extent, such measures were justifiable given the evident need to limit the spread of the disease and for its rapid eradication which would, of course, also be of benefit both to the tourism industry and a government facing an imminent general election. However, when it became evident that the outbreak was more serious and widespread than initially anticipated, subsequent policy decisions, in particular the eventual rejection of a vaccination strategy, served to confirm the view that the government was motivated primarily by the political (though economically unjustifiable) desire to retain the country's 'disease-free' status – hence maintaining its ability to export livestock and livestock products under the conditions of the 1993 GATT agreement (Midmore, 2001). As a result, the consequences for the tourism industry were, arguably, more severe that might have otherwise been the case. In particular:

- the policy of burning slaughtered animals and the backlog in disposing of carcasses created an image of the countryside that not only acted as a deterrent to potential visitors but also, and not surprisingly, one that the media transmitted to domestic and international audiences, further depressing potential demand. Indeed, research undertaken by the English Tourism Council (ETC) found that 46 per cent of respondents were put off visiting the countryside because of the sight of the destruction and disposal of animals whilst 38 per cent indicated that they would avoid visiting the countryside because of the potential health risks associated with burning animal carcasses (ETC, 2001c); and
- the policy of 'contiguous' culling – that is, the slaughter of animals on farms neighbouring and within a three kilometre radius of confirmed cases in accordance with EU regulations – not only exacerbated the problem of disposal, but also, quite simply, 'emptied' the countryside in the worst affected areas. In other words, one of the primary attractions to visitors, a living, working countryside, no longer existed, with inevitable consequences on demand.

This suggests that the potential consequences were either overlooked or not anticipated by the government – indeed, it was stated by the Minister for Tourism that, over three weeks into the crisis, the government was 'still not sure of what the effect was going to be on the tourism industry' (CMSC, 2001: xv). Either way, the impacts on tourism were immediate, severe and widespread.

The FMD Crisis: Impacts on Tourism

Inevitably, it will be some time before the full costs of the FMD crisis will be known although recent research suggests that, as a result of the FMD crisis, overall tourism expenditure in the UK in 2001 fell by £7.7bn; moreover, it predicts that in the following three years, the impact will continue to be felt with tourist expenditure reduced by £5.2bn, £1.3bn and 0.6bn in 2002, 2003 and 2004 respectively (see Blake *et al.*, 2002).

Nevertheless, beyond the estimates of the overall economic impacts referred to above, a number of points deserve mention:

- within two weeks of the start of the outbreak, tourism businesses in the countryside were suffering a severe downturn, although not all areas were affected to the same extent. For example, total March bookings for accommodation in Cumbria, the worst affected region, were down by 58 per cent compared with the previous year, with the Lake District National Park experiencing a 75 per cent fall (CMSC, 2001: xiii). Overall, it has been estimated that the Cumbrian tourism industry lost around £230mn (a decrease of around 25 per cent) in 2001. Non-accommodation facilities, such as cafes, restaurants, pubs and attractions, also experienced a virtual collapse in business with the vital domestic day visitor market putting off visits to the countryside. In the north-east, a survey during April found that 28 per cent of all small businesses had suffered a significant downturn in business, the worst affected being the hospitality sector which, during March, experienced an average 64 per cent decrease in turnover (CRE, 2001);

- the cancellation of a number of high-profile sports events had a severe knock-on effect on tourism. For example, the cancellation of the Cheltenham National Hunt Festival was expected to cost the local tourism economy some £20mn in lost income;

- it was not only the direct rural tourism industry that suffered. During April, over 80 per cent of businesses trading in the outdoor leisure sector experienced a significant reduction in turnover, with one in five suffering a 40 per cent drop in business (ETC, 2001a). As a specific example, the equestrian industry was reported to be losing over £100mn a month, with over 1,800 riding centres making heavy losses and some closing (Brown, 2001);

- inevitably, the crisis has also resulted in significant (mostly short-term) job losses throughout the tourism and leisure sector; by mid-April, 1,000 jobs had been lost in Cumbria, 18 per cent of businesses in Northumbria had laid off staff and, nationally, one sixth of outdoor leisure retailers had laid off staff; and

- domestic tourism business in general also suffered. For example, a number of domestic, coach-based tour operators experienced a downturn in business, with one suffering an 80 per cent drop in tourism business (Bowling, 2001).

In short, there is no doubt that the overall cost of the outbreak for the tourism industry, and the rural economy as a whole, was far greater, more widespread and longer lasting than initially anticipated by the government.

Policy Responses

The British government has long adopted a 'hands off' approach towards tourism. Indeed, since the 1969 Development of Tourism Act, successive governments have sought to minimise intervention in tourism, generally seeking to create a favourable framework for tourism to flourish (DCMS, 1999). This would, perhaps, explain the fact that, despite the obvious impacts of the disease control measures on rural tourism, 'the public authorities failed initially to appreciate the severity of the crisis for tourism' (CMSC, 2001: xv) and, moreover, sought to distance themselves from responsibility for these impacts. Thus, it was the 'widespread and often unbalanced coverage of the foot and mouth outbreak by British and overseas media... [that] ...has left many potential visitors uncertain about the extent to which the British countryside and its attractions are open' (DCMS, 2001: 3). In other words, not only was the government unable to appreciate the consequences of its actions in 'closing' the countryside, but also it then sought to distance itself from responsibility and avoid blame for those consequences.

Once it became apparent that action was necessary to address the challenges facing the tourism industry, policy responses fell under three broad categories:

- *attracting visitors back to the countryside* – actions included a variety of marketing efforts, including telephone 'hot lines'; the creation of web sites such as the British Tourist Authority's (BTA) 'openbritain.gov.uk'; political intervention in the form of overseas visits to promote Britain; and, the holding of a 'World Travel Leaders' summit. At the same time, additional funding was provided to both the BTA and English Tourism Council (ETC) to support immediate and longer-term campaigns to re-market Britain and the British countryside. A total of £14.2mn was allocated to the BTA and just £3.8mn to the ETC, figures not only well below what the two organisations considered they required for marketing purposes but, significantly, also which placed the emphasis on rebuilding the international tourist market. Rural tourism, however, depends primarily on the domestic market; typically, only around 6 per cent of overseas tourists visit the British countryside;
- *increasing accessibility to the countryside* – once the spread of FMD appeared to be under control, the government encouraged relevant organisations to re-open footpaths and areas of open countryside;
- *supporting businesses affected by the outbreak* – following the establishment of a Rural Task Force on 14 March, a number of measures were announced for local rural tourism businesses, including the ability to:

    - claim rate relief (from the local authority);
    - apply to the Inland Revenue and Customs & Excise for the deferral of tax and national insurance payments;
    - apply to bank managers for extended lines of credit or capital repayment holidays;

- apply for loans, guaranteed by the government, of up to £250,000 on a Small Business Service helpline;
- seek advice on employment issues from local job centres.

It is these latter measures that proved to be most controversial, being widely considered a confusing and insubstantial package of support. Not only were businesses expected to seek advice and financial assistance from a variety of sources, but the support was short term and principally on a deferral or loan basis. At the same time, a relatively complex bureaucratic procedure was required to secure support whilst, for many businesses, a loan at a relatively high interest rate (8.5 per cent) was of little value when they had little or no prospect of being able to repay it (CMSC, 2001).

More controversially, however, the support offered to rural tourism businesses contrasted starkly with the assistance and compensation provided to the agricultural sector which contributes significantly less to the national economy than tourism. As already noted, government policy with respect to controlling the outbreak of FMD – in particular, widespread culling as opposed to vaccination – was largely based on economic grounds. Two points should be emphasised:

- it was estimated that the potential value of livestock exports lost as a result of a vaccination policy, including the actual cost of vaccination itself, would have been no more than £315mn in 2001 (Midmore, 2001). Against this, the costs of the extended cull and consequential losses for farmers amounted to an estimated £600mn, although compensation and 'clean-up' costs resulted in an overall bill of about £2bn. Thus, on a simple economic basis, the policy of vaccination, thereby avoiding the widespread closure of the countryside and the decrease in tourism resulting from images of the cull and disposal process, would have been more cost effective; and
- the value of tourism to the rural and national economy is far greater than that of agriculture. In 2000, total tourism activity was estimated to have generated £64bn, four times as much as the total agricultural sector (CMSC, 2001; ETC, 2001b), and now accounts for some 7 per cent of national employment, compared with the 1.5 per cent contributed by agriculture. In the specific context of the countryside, tourism is worth some £12bn annually, and employs around 380,000 people in 25,000 small businesses. In some areas, particularly peripheral rural areas (not coincidentally, those worst affected by FMD), tourism has become the mainstay of the local economy and, of particular relevance here, 'rural tourism generates almost twice as much revenue as livestock farming' in Britain (CMSC, 2001: vii). Table 3.2 summarises the economic comparison between agriculture and tourism.

Table 3.1 Tourism and agriculture in Britain compared

	Agriculture (all)	Tourism
Revenue per annum	£15.3bn (total) £7.3bn (livestock & livestock products)	£64bn
GDP	1%	4%
Foreign exchange	£8.4bn (total) £1.0bn (livestock/dairy)	£12.5bn
Employment	1.5 % of workforce	7% of workforce
Tax contribution	£88m	£1.5bn
Growth rate: 1996-1999	-21% in revenue	+26% in revenue

Source: adapted from ETC (2001b)

Thus, not only is tourism is of greater economic significance to both the rural economy (compared with livestock farming) and the national economy, but it also generates significantly higher export earnings, it employs a larger proportion of the workforce and it remains a growth industry. However, this was not reflected in the government's response to the FMD crisis. What, then, are the implications for sustainable rural tourism development policy?

Implications for Rural Tourism Development

In one sense, the FMD crisis had a positive impact in that it brought (though, arguably, only temporarily) the value, scope and importance of rural tourism to the attention of both the government and the general public; the 'hidden giant' (CMSC, 2001) of the rural economy had at last been recognised. Indeed, the crisis demonstrated that rural tourism is not simply an alternative or extra source of income for farmers, but a diverse activity that directly and indirectly supports a wide range of industries and businesses. At the same time, the complex inter-relationship between rural tourism, agriculture, and the wider tourism industry as a whole has also become evident.

Conversely, the crisis has also served to highlight the inherent 'unsustainability' of the industry. For tourism to make an effective contribution to sustainable rural development it should, as this chapter has suggested, follow the governance model of planning and control. However, the policy responses to the outbreak outlined here demonstrated not only that ultimate power and control remains, of course, with central government, but also that the continuing health of the rural tourism industry is dependent upon the nature of the government's policies.

In the present context, those policies revealed, at the national level, a lack of recognition and understanding of the demand for and supply of rural tourism and the extent to which it is not only an integral element but also, arguably, the linchpin of rural economy and society. They also suggested that it has yet to be recognised that the countryside has undergone a fundamental socio-economic transformation. Agriculture, though still the dominant land use, has become less economically significant; tourism, conversely, is no longer simply an activity that supports or maintains the farming industry. Not only is tourism the more powerful economic force but, in a sense, farming supports the rural tourism industry by providing and maintaining one of the primary visual and cultural attractions of the countryside. Therefore, whilst it was and remains essential to support those farmers directly affected by the FMD outbreak, such support should not, arguably, be directed towards maintaining farming as an economic end in itself, but towards maintaining the farming community's contribution to the physical and cultural character of the countryside.

Inevitably, of course, agricultural policy, within Europe at least, remains tied into the European Union's Common Agricultural Policy. Nevertheless, the FMD crisis demonstrated that rural policy and tourism policy should be synonymous; such is the diversity of business supported by rural tourism, and such is tourism's contribution to conservation and the socio-economic regeneration of many rural areas, that tourism should become the focus of rural development policy. Thus, for rural tourism to become sustainable, for it to continue to contribute to sustainable rural regeneration and development, it is essential that it is given the national recognition and support it deserves. In other words, whilst the 'typical' elements of sustainable tourism policies, such as the localisation of labour, the sourcing of goods and service and so on, are applicable at a local level, the longer term contribution of tourism to sustainable rural development can only be achieved if it is given policy and financial support at the national level.

Note

1   At the same time, of course, a fundamental principle of sustainability is an holistic approach, whereby all forms of developmental activity should be considered within national developmental objectives and the global economic, social and environmental context.

References

Blake, A., Sinclair, M.T. and Sugiyarto, G. (2002), 'Quantifying the Impact of Foot and Mouth Disease on Tourism and the UK Economy' <www.nottingham.ac.uk/ttri/FMD-paper 4 .pdf>.

Bowling, K. (2001), 'Domestic Market Counts Cost of Countryside Crisis', *Travel Weekly*, 4 June, No. 1570, p. 2.

Brown, D. (2001), 'Rural Businesses Demand £12bn to Ease the Misery', *Daily Telegraph*, 10 May, p. 6.

Butler, R. (1998), 'Sustainable Tourism – Looking Backwards in Order to Progress?', in C.M. Hall and A. Lew (eds), *Sustainable Tourism: A Geographical Perspective*, Longman, Harlow, pp. 25-34.

Cavaco, C. (1995), 'Rural Tourism: The Creation of New Tourist Spaces', in A. Montanari and A. Williams (eds), *European Tourism: Regions, Spaces and Restructuring*, John Wiley & Sons, Chichester, pp. 129-49.

CMSC (Culture, Media and Sport Committee) (2001), *Tourism – The Hidden Giant – and Foot and Mouth, Fourth Report*, vol. 1, The Stationery Office, London.

Countryside Commission (1995), *Sustainable Rural Tourism: Opportunities for Local Action*, Countryside Commission, Cheltenham.

CRE (Centre for Rural Economy) (2001), *The Impact of the Foot and Mouth Crisis on Rural Firms: A Survey of Microbusinesses in the North East of England*, University of Newcastle, Newcastle-upon-Tyne.

DCMS (Department of Culture, Media and Sport) (1999), *Tomorrow's Tourism: A Growth Industry for the New Millennium*, DCMS, London.

DCMS (Department of Culture, Media and Sport) (2001), *National Tourism Recovery Strategy*, DCMS, London <www.culture.gov.uk/tourism/index.html>.

ETC (English Tourism Council) (2001a), 'Tourism Recovery: Foot and Mouth Campaign', <www.englishtourism.org.uk>.

ETC (English Tourism Council) (2001b), 'Foot and Mouth Disease and Tourism: Update from ETC', <www.englishtourism.org.uk>.

ETC (English Tourism Council) (2001c), 'Attitudes to Foot and Mouth in the Countryside', <ww.englishtourism.org.uk>.

ETC/CA (English Tourism Council/Countryside Agency) (2001), 'Working for the Countryside: A Strategy for Rural Tourism in England 2001-2005', London.

Galtung, J. (1986), 'Towards a New Economics: On the Theory and Practice of Self-reliance', in P. Ekins (ed.), *The Living Economy: A New Economy in the Making*, Routledge, London, pp. 97-109.

Garrod, B. and Fyall, A. (1998), 'Beyond the Rhetoric of Sustainable Tourism?', *Tourism Management*, vol. 19, pp. 199-212.

Godfrey, K. (1996), 'Towards Sustainability? Tourism in the Republic of Cyprus', in L. Harrison and W. Husbands (eds), *Practising Responsible Tourism: International Case Studies in Tourism Planning, Policy and Development*, John Wiley & Sons, Chichester, pp. 58-79.

Goodwin, M. (1998), 'The Governance of Rural Areas: Some Emerging Research Issues and Agendas', *Journal of Rural Studies*, vol. 14, pp. 5-12.

Harrison, C. (1991), *Countryside Recreation in a Changing Society*, TMS Partnership, London.

Hoggart, K., Buller, H. and Black, R. (1995), *Rural Europe: Identity and Change*, Arnold, London.

Hunter, C. (1995), 'On the Need to Re-conceptualise Sustainable Tourism Development', *Journal of Sustainable Tourism*, vol. 3, pp. 155-65.

Marsden, T. and Murdoch, J. (1998) 'The Shifting Nature of Rural Governance and Community Participation', *Journal of Rural Studies*, vol. 14, pp. 1-4.

Midmore, P. (2001), 'The 2001 Foot and Mouth Outbreak: Economic Arguments Against an Extended Cull', <www.sheepdrove.com/fam.html>.

Newby, H. (1985), *Green and Pleasant Land? Social Change in Rural England*, Wildwood House, London.

Oppermann, M. (1996), 'Rural tourism in Southern Germany', *Annals of Tourism Research*, vol. 23, pp. 86-102.

Roberts, L. and Hall, D. (2001), *Rural Tourism and Recreation: Principles to Practice*, CABI Publishing, Wallingford.

Sharpley, R. (2000), 'Tourism and Sustainable Development: Exploring the Theoretical Divide', *Journal of Sustainable Tourism*, vol. 8, pp. 1-19.

Sharpley, R. (2002a), 'Tourism – a Vehicle for Development?', in R. Sharpley and D. Telfer (eds), *Tourism and Development: Concepts and Issues*, Channel View Publications, Clevedon, pp. 11-34.

Sharpley, R. (2002b), 'Sustainability: A Barrier to Tourism Development?', in R. Sharpley and D. Telfer (eds), *Tourism and Development: Concepts and Issues*, Channel View Publications, Clevedon, pp. 319-37.

Sharpley, R. and Craven, B. (2001), 'The 2001 Foot and Mouth Crisis – Rural Economy and Tourism Policy Implications: A Comment, *Current Issues in Tourism*, vol. 4, pp. 527-37.

Shoard, M. (1999), *A Right to Roam?*, Oxford University Press, Oxford.

Southgate, C. and Sharpley, R. (2002), 'Tourism, Development and the Environment', in R. Sharpley and D. Telfer (eds), *Tourism and Development: Concepts and Issues*, Channel View Publications, Clevedon, pp. 231-62.

Stoker, G. (1997), 'Public-private Partnerships and Urban Governance', in G. Stoker (ed.), *Partners in Urban Governance: European and American Experiences*, Macmillan, Basingstoke, pp. 1-21.

Stone, P. (2000), *Countryside Ideologies and Rural Governance: The Case Study of the Lake District*, unpublished MSc thesis, University of Northumbria, Newcastle-upon-Tyne.

Streeten, P. (1977), 'The Basic Features of a Basic Needs Approach to Development', *International Development Review*, vol. 3, pp. 8-16.

Wall, G. (1997), 'Sustainable Tourism – Unsustainable Development', in S. Wahab and J. Pigram (eds), *Tourism, Development and Growth: the Challenge of Sustainability*, Routledge, London, pp. 33-49.

WCED (1987), *Our Common Future*, World Commission on Environment and Development, Oxford University Press, New York.

WTO/WTTC (1996), *Agenda 21 for the Travel and Tourism Industry: Towards Environmentally Sustainable Development*, World Tourism Organisation/ World Travel and Tourism Council, Madrid.

Yearley, S. (1992), *The Green Case: A Sociology of Environmental Issues, Arguments and Politics*, Routledge, London.

Yeoman, J. (2000), 'Achieving Sustainable Tourism: a Paradigmatic Perspective', in M. Robinson (ed.), *Reflections on International Tourism: Environmental Management and Pathways to Sustainable Tourism*, Business Education Publishers, Sunderland, pp. 311-26.

Chapter 4

What is Managed when Managing Rural Tourism? The Case of Denmark

Anders Sørensen and Per-Åke Nilsson

Introduction

In recent years, rural tourism has attracted a steadily increasing level of research attention, stimulated by two factors. First, tourism demand for rural areas is growing (Cavaco, 1995; Dernoi, 1983; Fleischer and Pizam, 1997; Hummelbrunner and Miglbauer, 1994; Lane, 1994). Second, rural tourism has arrived on the political-economic agenda, the hope being that tourism business can alleviate the consequences of a decline in traditional means of rural employment (Butler, Hall, and Jenkins, 1998; Cavaco 1995; Fleischer and Pizam 1997; Gannon, 1994; Hjalager, 1996; Lane 1994; Luloff et al., 1994; Nilsson, 1996; 1999; Page and Getz, 1997b; Sharpley and Sharpley, 1997; Sørensen, 1997; Weaver, 1997).

Supply side issues of rural tourism have attracted particular attention. This is not only because of the importance ascribed to the latter of the above two factors; it also compounds a general inclination within tourism research to view tourism from a supply side viewpoint. This is not to say that tourism research has neglected the tourists, but mostly tourists and their actions are only interesting in so far as they affect the supply side or impact upon receiving areas. Owing to this, tourism research is somewhat lacking in concepts and models that encourage a more subtle insight into tourism demand and consumption.

This is also the case for rural tourism. For despite growing rural tourism research, we still have but superficial insight into rural tourism as viewed from the tourists' point of view. Knowledge is scarce on the patterns of consumption among visitors to rural tourism facilities, or on the spatial patterns of tourism in which rural tourism facilities are included. Are rural tourists rural throughout the holiday? If not, what do they chain together with the rural elements? What do they consume when they are elsewhere than the site where the researchers do the surveying? More robust knowledge on such matters will be a valuable foundation for more general social scientific reflections on the consumption of rurality.

The impact of such matters reaches beyond academia and into issues of policy-making and resource- and business-management. In a Danish context, the relationship between rural tourism and rural recreation illustrates this. None of these notions are unambiguous, and while distinct in societal legitimacy, recreation and tourism nevertheless overlap in their activities in, and consumption of, the rural landscape. A resident population's recreational use of an area draws on the same resources as tourist visitors and recreational visitors. Usually, the resident population has the advantage in this contest. Not only do they benefit from proximity, but they may also enjoy privileges in terms of access or use, privileges that may be based on ownership, locally distributed rights of use, or a network that enables access not allowed to 'outsiders'. Conversely, many non-local recreationists or tourists can be viewed as underprivileged in recreational terms (Nilsson, 2002a).

Thus, two polarities are easily identified: one between local and non-local users of rural facilities, and another between tourism use and recreational use. However, the rigidity of these polarities is dissolving in contemporary society. In many regions, the rural-urban divide is much less distinct than previously. Both academically and politically, there is widespread recognition of these changes, and thereby also of the disappearance of a traditional local/non-local polarity.

The other polarity, however, the one between recreation and tourism in rural areas, seems, at the ideological and political level, largely unaffected by the societal changes taking place. In short, these changes can be described as changing patterns of leisure consumption and a de-differentiation of spheres of consumption, whether touristic, recreational, or experiential (Rojek, 1993, 1995). Yet, in a Danish context this de-differentiation has only to a limited degree affected public debate, where two autonomous discourses have been maintained.

One discourse is concerned with recreation and public access to recreational opportunities and facilities. This discourse is founded on societal morals concerned with assumed positive effects of recreational activity, the re-creation and upholding of the individual as a productive member of society. Within this discourse, the individual is seen as deserving the right to recreational access, for the good of self and society. The other discourse is concerned with tourism – but not with supposed positive effects for the individual tourist. The discourse is concerned with tourism as commerce: economic effects and local consequences. In this way, tourism and recreation in rural areas, although utilising the same natural resources, are to a large degree debated separately, and often in contradictory terms. This matter is clearly something worth looking into.

This chapter examines the conception of rural tourism as it appears in the research literature. This is compared with qualitative data on tourism and recreation in rural Denmark in order to consider the merit of such conceptions in a Danish context. It is found that there is more rural tourism than rural tourists in Denmark, in the sense that most patrons of rural tourism facilities did not meet the conception of rural tourists that the research literature implicitly conveys. The findings on this point to a two-edged problem: on the one hand, due to a perceived negligibility of rural tourism in Denmark, it is left out from any significant influence on the public debate on the priorities and management of rural resources. On the other hand, this perceived negligibility is sustained by academic conception of rural tourism, which stresses matters of distinction, rather than matters of phenomenological correspondence between tourism, recreation, residence, and other means of appropriation of the rural which is found in postmodern, de-differentiated consumption.

Conceiving of Rural Tourism and Recreation

Although apparently unambiguous, the term rural is nevertheless problematic, and critical examination of the concept of rural seems to have had only little impact in social science. There is an awareness of this within tourism research. Thus, quoting Hoggart, Page and Getz argue that: *"the undifferentiated use of 'rural' in a research context is detrimental to the advancement of social theory"* - since the term rural is obfuscatory due to intra-rural differences and urban-rural similarities' (Page and Getz, 1997a: 5). Rural tourism research, it would seem, offers a privileged platform from which to consider a rethinking and re-conceptualisation of the concept of rural, since rural tourism in itself displays ample evidence that 'rural' is more than simply a matter of non-urban.

Basically, one must agree with Sharpley and Sharpley (1997: 5) that no definition of rural tourism is generally accepted. However, while it would be futile to construct yet another definition of rural tourism, it is useful to contemplate some of the depictions and conceptions of rural tourism present in tourism research and in public discourse.

The EU tends to view all tourism in non-urban areas as rural tourism (Nilsson, 1998). However, while perhaps necessary for rudimentary descriptive administrative purposes, such a view of rural tourism has no analytical value.

Oppermann (1996) separates 'wilderness tourism', which stands for tourism in uncultivated areas, from 'rural tourism', which is tourism in a non-urban area with human activity and permanent human habitation; in other words, agrarian areas. In this way, Oppermann employs an intuitively appealing distinction between 'nature' and 'culture' where the tilled land is perceived as 'culture', as it is in most societies around the world. However, being based on geographic territory, the definition stresses locality but discounts activity or perception.

Page and Getz (1997a: 7) and Bramwell (1994: 2) address these matters by asking whether rural tourism is something special, something distinct, brought about by the physical, cultural and social setting, or whether it is simply tourism as any other tourism, this time taking place in rural surroundings? In other words, is it relevant to use the term 'rural tourism', or is it more fitting to use the less significant term 'tourism in rural areas'? If the former is the case, it implies that there is something beyond location that makes rural tourism distinct.

Actual exploits undertaken, however, are not enough to mark out rural tourism, since many activities or impressions can be experienced by other types of users, e.g. day-trippers or local inhabitants. Page and Getz (1997a) find that rural tourism is underestimated in tourism research, due to the fact that it is often viewed as recreation rather than business. This points towards an interesting distinction between users' activities, which are viewed as recreation, and the suppliers' attractions and undertakings, which are viewed as tourism business. This distinction is also evident in the way that Oppermann distinguishes between rural tourists and rural recreationists by the formers' use of rural accommodation (Oppermann, 1996: 88).

Thus, locality and accommodation seem to be the distinguishing factors. However, a perceptual or impressionistic dimension is also present. It seems that those activities and settings that commonly are thought to be encompassed by the term 'rural tourism', are somehow 'in agreement' with the landscape: it employs the rural landscape and not just the territory. Thus, a holiday park or a theme park located in a rural area is not thought of as rural tourism if it employs only the physical territory and not any inherent or attributed characteristics of the rural landscape. Conversely, an open-air museum or a 'medieval experience' attraction includes elements of rurality by its use of rural landscape aesthetics.

Nonetheless, many would reject classifying such attractions as rural tourism. The rejection would be based on matters of volume. Lane (1994) argues that rural tourism must be rural in function and scale, based on local traditions with local roots, and based on perceptions of what is rural. From this standpoint, the concept of rural tourism would not include large-scale business ventures.

The issue of rural in function and scale also shows itself to be significant as regards matters of delimiting rural tourism geographically. For the notion of rural in the concept of rural tourism does not always signify a distinction between various categories of non-urban landscape: the cultivated landscape, the nurtured landscape, the nurtured nature, and the untended nature. It seems that almost all categories of non-urban landscape can be captured within the concept of rural tourism – except the pleasure beach. Both popularly and academically, the beach is conventionally understood as an area of deliberately hedonistic pleasure-seeking (Cohen, 1982; Lencek and Bosker, 1998; Shields, 1990; Wagner, 1977), and as a liminal zone of hedonism, it is in opposition to the underlying connotations of authenticity and serious pleasure, which is contained in the notion of rural tourism. Rural tourism is thought of as the opposite of beach tourism since beach tourism is viewed as large-scale mass tourism with a collective tourism attitude or gaze (cf. Urry, 1990) whereas rural tourism is viewed as an individualistic venture. Thereby, the notion of rural tourism seems also to contain an element of tourism aesthetics, through a distinction between 'respectable' tourism and the tourism of the masses. Rural tourism is precluded from being a mass tourism phenomenon, for as soon as the masses arrive, it is no longer rural tourism!

In this way, the notion of rural tourism shares the same origins as the notion of alternative tourism. In both cases, formal definitions can be constructed and argued but behind these definitions lie certain moral assumptions about undesired modes of tourism. This causes both rural tourism and alternative tourism to be conceived by means of negation: *by what it is not*. In this way, rural tourism is implicitly conceived of as non-urban, i.e. rural; non-coastal, i.e. non-hedonistic; non-mass, i.e. small-scale; non-packaged, i.e. individualistic. In this way, conceptions of both alternative and rural tourism draw on a patronising critique of the proliferation of tourism. By the rhetoric technique of associating it with the critique of mass tourism, rural tourism, as alternative tourism, is attributed with positive connotations since it comes close to being perceived as an anti-tourism tourism (cf. Cohen, 1989). Thus, the attraction of the rural not only consists of images and attractions of the rural, but also of impressions of a 'counter-mode' of tourism.

Interestingly, a similar distinction is found in the prevalent demarcation between tourism and recreation. Rural recreation is normally understood as a possibility for ordinary people to gain access to recreational areas. Rural tourism on the other hand, is viewed as a commercial transaction, which enables people to visit a place, get service and pay for it. This division causes great boundary problems (cf. Emmelin, 1996). Both forms can coexist and overlap, but recreation and tourism in rural areas continue to be vastly different in their social legitimacy.

In the Field

Fieldwork was conducted in three Danish rural municipalities: Rangstrup in inland South Jutland, Ryslinge in the heart of Funen, and coastal Stevns in South-East Zealand. Rangstrup offers plenty of accommodation, mainly holiday cottages and campsites, but almost no named sights or attractions. When the authors inquired at the local tourism information, the attractions suggested were all located outside the municipality. In contrast, Ryslinge municipality has almost no commercial accommodation but boasts the mega-attraction of Egeskov castle, complete with moat, extensive gardens, museums and exhibitions, located in a romantic rural landscape, and attracting more than 250,000 visitors in the season. The third municipality, Stevns, falls between the two above extremes in that it displays several attractions, and a reasonable capacity and variety of commercial accommodation, camp sites, hotels and holiday cottages. None of the three municipalities offered many opportunities for farm-stays or the like.

The municipalities posses the recreational opportunities that private and public forests offer, within the restrictions that public use of such areas is subjected to in Denmark. Furthermore, Rangstrup has golf courses and put-and-take fishing, while Stevns, the only littoral municipality of the three, has fine beaches. All three municipalities are criss-crossed by regional and national bicycle routes, while a national scenic route designed for motoring passes through Ryslinge and Stevns, but not Rangstrup. In total, the three municipalities reflect the diversified distribution of facilities for rural tourism and recreation in Denmark.

Although not robust enough to justify sweeping generalisations, the fieldwork data do display an interesting uniformity in terms of a *lack* of focus on the rural and rurality in the informants' utterances, acts, and holiday arrangements.

During fieldwork, the authors used the diversity of accommodation found in the three municipalities, in order to facilitate contact with a wide range of tourists accommodated in the areas. Although ethnographically founded, data collection at accommodation facilities was based on interviews and conversations rather than participant-observation. Additionally, interviews and participant-observation took place at attractions and locations of various activities in the rural landscape. In many ways, data-producing opportunities, both observation- and interaction-based, were more readily accessible at such places. However, the data represented both tourists' and recreationists' use of activities and attractions.

While exhibiting the variation of attractions and activities in the Danish rural landscape, the three municipalities also demonstrate the vast overlap between touristic and recreational utilisation of facilities. For all the attractions and activities, including nature/forests in the three municipalities, are being used for both tourism and leisure/recreation purposes. Thus, rural tourists could not be distinguished from other leisure users of rural areas only by means of activities.

Nor could the use of rurally located accommodation serve to identify rural tourists or a distinct rural tourism. Whether accommodated on campsites, holiday cottages, farm-stays or B&Bs, almost all of the patrons used the rural accommodation as a base camp for activities that included both urban tourism and beach tourism. The short distances in Denmark (nowhere is more than 65km to a beach), allow for a wide range of both urban and beach tourism activities within day-trip distance from a rural base camp. The proximity of non-rural elements thus enables the non-rural elements to be prominently present in the holiday consumption of rurally accommodated tourists. But equally important, the choice of rural accommodation did not seem to be deeply motivated by perceptions of rurality. A few examples illustrate the point.

In Rangstrup municipality, the authors interviewed both owner-users and renters of holiday cottages. They all stated that they were attracted to the area because it implied going to the countryside for holiday. Yet, when describing their actual holiday activities, they included a number of obviously non-rural elements, such as theme parks (Legoland), beaches, historic cities, and cross-border shopping trips to Germany. Certainly, many other activities included the use of the area's rural recreational facilities, like walking the dog in the nearby forest, but all of the activities would be possible to include in the same persons' everyday life, even in an urban area.

Another example was a couple on holiday, encountered at a farm-stay in Stevns municipality. Being rural residents themselves, they stayed at six different farm-stays during their two-week domestic holiday. The reason behind their preferences, however, did not hinge on matters of rurality, but matters of scale. The wife stated: 'I much prefer places like this. In a big place, it's very formal, you feel like a stranger, whereas in this place you get a chat with the landlady.' For this couple, the attractiveness of the farm-stays was based on perceived familiarity, not perceived rurality.

The authors stayed at different farm-stays or farm based B&Bs. Interestingly, those providers who most clearly marketed their product as farm-stays, were also those with the least actual farming attached. At these places, the 'farm' dimension was a matter of aesthetics rather than production. Yet, even at such accommodation, which otherwise would be expected to be the place where, if anywhere, one could encounter the 'dedicated' rural tourist, such types could not be identified among the patrons. None of the visitors went to the countryside in order to fully 'go rural', experientially or otherwise; nor did they aim for small-scale accommodation facilities in the belief that 'small is rural'. A large share of the customers were *en route* on a multi-destination holiday and did not stay more than one or two nights.

In total, tourists in rural Denmark do *not* display any tendency to overly romanticise or 'idyllicise' the countryside in their explanations of why rural areas were included in their holiday. The explanations did *not* display any strong emphasis on a pursuit of the authentic in the countryside. The presence of such attachments was located at a deeper structural level in the sense that most informants were city-dwellers and wanted to experience rural nature, beauty and tranquility. The perceived urban-rural inversion present in such perceptions, and the pursuit of this inversion (rather than another) seems to be founded on an ideology or imagery of the rural which the authors suspect is widespread in Western society (Nilsson, 2002b; Sørensen 1997). It is at this level, rather than at the immediate level of consumption or articulation, that the idea of 'the countryside' is constructed, and becomes influential on the individual's holiday choices (Sørensen and Nilsson, 2000).

Reflections

The findings presented above did not identify any clear differences between tourist and recreational use of the rural landscape. Furthermore, the lack of differentiation in actual usage is repeated in terms of perceived differences. The findings indicate that in the minds of the average Danish citizens, there is no obvious disagreement between tourism use and recreational use of the rural agrarian landscape.

This does not mean that there are no conflicts. Rural Denmark displays numerous types of clashes of interest in which recreation or tourism is a focal point in the contestation. These include dual-use antagonism, local annoyance with non-locals' use of recreational resources, littoral site owners' discontent with the building restrictions imposed by the coastal protection act, in-migrated residents' protests against local tourism business expansion plans, etc. Such clashes of interest are of course present in rural Denmark. However, certain overriding factors influence and control the articulation and character of such conflicts.

One such factor is the way in which the rural landscape is appropriated and consumed through recreation and tourism. In Denmark, the rural landscape is predominantly enjoyed visually, as an aesthetic experience. While roaming behaviour is found (e.g. hunting, horse riding, off-road bicycling), such activities are predominantly performed by local residents. Furthermore, not only is the use of the terrain for such activities constrained by legal means, but these activities are also controlled by behavioural morals, in the sense that Danes' domestic nature behaviour is remarkably controlled and disciplined. Danes stick to the trails in the forest and in the open landscape; they do not trespass on farm land, not even pastures or fallow land, and camping outside recognised facilities is almost unheard of.

The Danish Nature Protection Act places severe restrictions on 'disorderly' camping outside recognised facilities, and other parts of legislation controls and restricts public access to and behaviour on private and public land. The Danish public in general accepts these restrictions. While they admire and envy the public access and right to roam found in the other Nordic countries, and while many Danes utilise the right to roam when on holiday in the other Nordic countries, this has not led to any strong demands being raised for similar public rights of access and use in non-littoral rural Denmark. Popularly, this is explained by referring to population density, thereby implying that restrictions are necessary in order to protect Danish nature from excessive exploitation. While this explanation is very alluring when comparing Denmark with the other Nordic countries, it is also convenient for several actors in the public debate in Denmark.

Denmark is a very intensively farmed country whose economy throughout industrialisation depended on the export of agricultural produce. Historically, the combination of a very intensive land use and a national economy which for a long time depended on the export of agricultural produce, has enabled agricultural bodies to wield much influence and to set the agenda for the use of the open landscape. They have not been in favour of any increase in the accessibility for ordinary people to penetrate the landscape. For nature protection agencies and organisations, it has been important to widen the interest of ordinary people for in nature, mostly from an educational point of view. But even here, the agenda set by the agriculture industry has been the frame within which such attempts have been formed.

Recently, traits of a more active and proactive attitude in the public can be discerned. Access has come more and more to be viewed as a personal and democratic right, and the monopoly held by the agrarian side setting the agenda for the use of nature has been challenged. Albeit still in rather modest forms, the issue of widened access for ordinary people to the rural landscape has been raised. Parallel to that, the opinion from the agriculture lobby that people do not desire any wider access to nature than what they presently have is undergoing increasing scrutiny. Still, the interrelated factors outlined above, combined with strict and complex zoning laws, mean that in Denmark, rural tourism is perceived as a matter of accommodation, since everything else connected to rural tourism is regulated and controlled as recreation. To be sure, business developments like golf courses or amusement parks do take place in rural areas, but such developments are not perceived as rural tourism developments, but rather as urban implants in the rural landscape.

Consequently, as rural tourism in Denmark therefore is viewed as a matter of commercial accommodation in non-littoral, non-urban areas, the tautological inference is strikingly near. For since, ostensibly, there is not much rural tourism, rural access is continuously viewed as something that should be administered according to recreational needs (perceived as educational and recreational, and nature-preserving) rather than tourism needs (perceived as commercial and exploitative and nature-destructive). Therefore, rural tourism is not left with much leverage in the public debate on the priorities and management of rural resources. It is 'out-delimited', so to speak, from any significant influence on this discourse.

Ironically, the upsurge in rural tourism research has not done much to alleviate this. If anything, academic conceptions of rural tourism as outlined previously, instead entrench the administrative marginalisation of rural tourism witnessed in Denmark, since the underlying conceptual focus is on matters of distinction (non-mass, non-urban, non-hedonistic, non-packaged) from conventional high-volume tourism, rather than matters of affiliation to other forms of consumption of the rural. This point is important, for the consumption of the rural seems to be growing but in ways which cannot always be classified as either tourism or recreation. The apparent distinction between tourism and recreation is contradicted by an increasing blurring of traditional spheres of consumption and activity. Previous distinctions (e.g. work and leisure, travel and residence, urban and rural) is being de-differentiated, at times radically, and in many ways this impacts upon the use, appropriation and consumption of the rural. Seen in this light, rural tourism is but one of many means of consumption of the rural within a Danish context.

Conclusion

While taking its point of departure in tourism research, this chapter has considered rural tourism and recreation practices in Denmark, as seen in the light of a programmatic analysis of conceptions of rural tourism in research and public discourse. One conclusion is abundantly clear: not only is it difficult to identify a distinct rural tourism in Denmark; but tourism and recreation are so overlapping in terms of actual activities in and use of the rural landscape in Denmark, that it necessarily has to be taken into account in actual resource management.

As it is, this is most likely not a problem as regards tourism/recreation businesses that exploit rural resources in the service of the leisure economy. From the point of view of most rural tourism entrepreneurs, the question 'what is rural tourism' is largely of no great concern (Page and Getz, 1997a). A customer is a customer, and whether scholars term them as rural tourists or otherwise is businesswise a matter of indifference.

Public resource management and policy practice, however, is another matter. It is likely to be of some consequence that while recreational use of rural resources, if used within limits set by agriculture and nature protection, is viewed as *morally and societally right*, tourism use of the same resources is viewed as *commercially sane*. Improved management of public recreational resources and facilities would be facilitated by a de-differentiation of the symbolic boundaries between tourism and recreation.

Interestingly, the encouragement for such a change in public and administrative perception is more likely to come from the playful incursion of the postmodern in public life, than from rational arguments or from rural tourism research findings. The difference between actual tourism and recreation practices in rural areas on the one hand, and the symbolic and ideological distinction between the two on the other is eminently modernist in its maintenance of distinct cultural spheres with distinct production of symbolic capital. However, this distinction seems to be eroding (Betz, 1992; Featherstone, 1991; Urry, 1995; Urry, 2000). As a consequence, the 'intimations of postmodernity' (Bauman, 1992) that are found in contemporary society might result in the actual overlap between tourism and recreation consumption of rurality and rural areas being recognised as such, and being acted upon in policy practice and resource management.

Acknowledgement

The research presented here is funded by the Danish Research Council's research programme: *The Agrarian Landscape in Denmark*.

References

Bauman, Z. (1992), *Intimations of Postmodernity*, Routledge, London.
Betz, H.G. (1992), 'Postmodernism and the New Middle Class', *Theory, Culture & Society*, vol. 9, pp. 93-114.
Bramwell, B. (1994), 'Rural Tourism and Sustainable Rural Tourism', *Journal of Sustainable Tourism*, vol. 2, pp. 1-6.
Butler, R.W., Hall, C.M., and Jenkins, J. M. (eds) (1998), *Tourism and Recreation in Rural Areas*, John Wiley & Sons, Chichester.
Cavaco, C. (1995), 'Rural Tourism: The Creation of New Tourist Spaces', in A. Montanari and A.M. Williams (eds), *European Tourism: Regions Spaces and Restructuring*, John Wiley & Sons, Chichester, pp. 127-49.
Cohen, E. (1982), 'Marginal Paradises: Bungalow Tourism on the Islands of Southern Thailand', *Annals of Tourism Research*, vol. 9, pp. 189-228.
Cohen, E. (1989), 'Alternative Tourism: A Critique', in T.V. Singh, H.L. Theuns, and F.M. Go (eds), *Towards Appropriate Tourism: The Case of Developing Countries*, Peter Lang, Frankfurt am Main, pp. 127-42.
Dernoi, L.A. (1983), 'Farm Tourism in Europe', *Tourism Management*, vol. 4, pp. 155-56.
Emmelin, L. (1996), *Natur, Turism och Förvaltning*, Department of Tourism Studies, Mid-Sweden University, Östersund.
Featherstone, M. (1991), *Consumer Culture and Postmodernism*, Sage, London.
Fleischer, A. and Pizam, A. (1997), 'Rural Tourism in Israel', *Tourism Management*, vol. 18, pp. 367-72.
Gannon, A. (1994), 'Rural Tourism as a Factor in Rural Community Economic Development for Economies in Transition', *Journal of Sustainable Tourism*, vol. 2, pp. 51-60.
Hjalager, A.M. (1996), 'Agricultural Diversification into Tourism: Evidence of a European Community Development Programme', *Tourism Management*, vol. 17, pp. 103-11.
Hummelbrunner, R. and Miglbauer, E. (1994), 'Tourism Promotion and Potential in Peripheral Areas: The Austrian Case', *Journal of Sustainable Tourism*, vol. 2, pp. 41-50.
Lane, B. (1994), 'What is Rural Tourism?', *Journal of Sustainable Tourism*, vol. 2, pp. 7-21.

Lencek, L. and Bosker, G. (1998), *The Beach: the History of Paradise on Earth*, Pimlico, London.

Luloff, A.E., Bridger, J.C., Graefe, A., Saylor, M., Martin, K., and Gitelson, R. (1994), 'Assessing Rural Tourism Efforts in the United States', *Annals of Tourism Research*, vol. 21, pp. 46-64.

Nilsson, P.Å. (1996), 'Jordbruk och Turism i Perifera Regioner', in S. Lundtorp (ed.), *Udkantsområder - Regional- og Turismeforskning på Bornholms Forskningscenter*, Research Centre of Bornholm, Nexø, pp. 251-84.

Nilsson, P.Å. (1998), *Bo på Lantgård - en Studie av Bondgårdsturism som Idé*, Research Centre of Bornholm, Nexø.

Nilsson, P.Å. (1999), 'Tourism's Role in a New Rural Policy for Peripheral Areas: The Case of Arjeplog', in F. Brown and D. Hall (eds), *Case Studies of Tourism in Peripheral Areas*, Research Centre of Bornholm, Nexø, pp. 157-80.

Nilsson, P.Å. (2002a), *Rekreation og Tilgængelighed i et Tætbefolket Område - Brug og Attituder i det Agrare Danmark*, Center for Regional- og Turismeforskning, Nexø.

Nilsson, P.Å. (2002b), 'Staying on Farms: An Ideological Background', *Annals of Tourism Research*, vol. 29, pp. 7-24.

Oppermann, M. (1996), 'Rural Tourism in Southern Germany', *Annals of Tourism Research*, vol. 23, pp. 86-102.

Page, S.J. and Getz, D. (1997a), 'The Business of Rural Tourism: International Perspectives', in S.J. Page and D. Getz (eds), *The Business of Rural Tourism: International Perspectives*, International Thomson Business Press, London, pp. 3-37.

Page, S.J. and Getz, D. (eds) (1997b), *The Business of Rural Tourism: International Perspectives*, International Thomson Business Press, London.

Rojek, C. (1993), *Ways of Escape*, Macmillan, London.

Rojek, C. (1995), *Decentring Leisure: Rethinking Leisure Theory*, Sage, London.

Sharpley, R. and Sharpley, J. (1997), *Rural Tourism: An Introduction*, International Thomson Business Press, London.

Shields, R. (1990), *Places on the Margin*, Routledge, London.

Sørensen, A. (1997), 'Turismens Agrare Landskab', in S. Andersen (ed.), *Det Agrare Landskab i Danmark: Foredrag fra Seminar 1*, Forskningsrådene/Danish Research Councils, København, pp. 128-36.

Sørensen, A. and Nilsson, P.Å. (2000), 'Virtual Rurality Versus Rural Reality – Contemplating the Attraction of the Rural', in *Hovedforedrag og Papers fra det 8. Nordiske Forskersymposium i Turisme, Alta, 18. - 21. November 1999*, Høgskolen i Finnmark, Alta, pp. 379-92.

Urry, J. (1990), *The Tourist Gaze*, Sage, London.

Urry, J. (1995), *Consuming Places*, Routledge, London.

Urry, J. (2000), *Sociology Beyond Societies: Mobilities for the Twenty-first Century*, Routledge, London.

Wagner, U. (1977), 'Out of Time and Place: Mass Tourism and Charter Trips', *Ethnos*, vol. 42, pp. 38-52.

Weaver, D.B. (1997), 'The Vacation Farm Sector in Saskatchewan: A Profile of Operations', *Tourism Management*, vol. 18, pp. 357-65.

PART 3
EXPERIENCE

Chapter 5

Encouraging Responsible Access to the Countryside

Lesley Roberts and Fiona Simpson

Introduction

The Land Reform (Scotland) Act 2003 introduces a new right of responsible access in Scotland. The legislation aims to reflect a longstanding tradition of tolerance and respect towards the need for people to move freely within the countryside. Prior to the new Act, access rights in Scotland constituted a complex combination of common and statute law; apart from routes explicitly defined as rights of way in common law, countryside users required express or implied consent to access which could be revoked by landowners at their discretion. This created a lack of clarity resulting in confusion and dispute where recreational users attempted to exercise rights they did not possess.

The legislation represents a bold statement by a new Scottish Parliament for it stirred entrenched attitudes and emotional responses to issues of land ownership and control that many see as one of the country's defining characteristics. Over 85 per cent of land in Scotland is in private ownership, much of it in the hands of non-Scottish and absentee landlords. Initially, the Land Reform Bill proposed the right to responsible access to all rural land (including enclosed land), further challenging understanding of the role of Scotland's working countryside at a time when changes to EU Common Agricultural Policy are placing strains on farming and landowning communities.[1]

Although sentiment underpinning the legislation has been broadly welcomed, its passage through public consultation and parliamentary debate brought into sharp relief the varying perceptions of countryside access, highlighting the need to achieve a balance between the aspirations of recreational users and the needs of landowners and managers, and it was feared that the opportunity to establish what may be perceived as a right to roam may be lost as result of lobbying by concerned land owners and managers (Shoard, 2000). During the consultation process, which elicited some 3,500 submissions, one key point of consensus was the emphasis on *responsible* access. In parallel with the preparation of the Act, Scottish Natural Heritage (SNH) was commissioned to prepare and promote a new Scottish Outdoor Access Code to provide advice to users, land managers and local authorities on the definition of the term 'responsible access'. A draft Code was published for consultation alongside the original Bill, and was updated following the consultation period in December 2001. The Act received Royal Assent in February 2003.

Set within the context of access management in Scotland, this chapter builds on the debate on access rights and progresses to an analysis of the context within which 'responsible access' might be achieved. By looking forward to the implementation of the Outdoor Access Code, the discussion aims to expose the complexities inherent in seeking to influence user behaviour. The chapter also outlines the means by which interpretative materials may be helpful in the implementation of the Code as a means of ensuring that the new Land Reform Act helps to ameliorate, rather than to exacerbate, continuing potential conflicts between users and land managers.

Land Management Responses to the New Legislation

The new legislation is likely to have a direct impact on relationships between land managers, recreational users of the countryside and more generally local communities, particularly on the urban fringe. The new provisions are potentially divisive, requiring careful and balanced application if they are to avoid widening the gap in perceptions and values that tend to exist between these groups. Amongst these differences, the land managers' view that the countryside is primarily a living and working environment is fundamentally at odds with the consumption of the countryside as a tourism and recreational resource, and this has been evident in land managers' responses to the legislation as it emerged and was amended over time.

Recreation and tourism in rural parts of Scotland continue to grow in popularity, and it remains the case that these need to be better managed in the interests of all parties. The new legislation provides an opportunity to extend understanding of access issues and to ensure that broader 'communities of interest' become involved in the management of resources (Simpson, 2001: 87). The extent to which this can be achieved, however, was called into question throughout the period of consultation on the (then) proposed legislation. In particular, the withdrawal of the National Farmers' Union for Scotland from the National Access Forum (the body responsible for the search for resolution through debate) raised concerns about the extent to which the land owners and managers can play a constructive role in partnership with other countryside user groups such as the Ramblers' Association.

Land managers share a number of common concerns about access, many of which require attention if the new legislation is to avoid further polarisation of views within the ongoing debate. Farmers in urban fringe areas feel particularly vulnerable to increased access which, if not exercised responsibly, may lead to damage and considerable cost. If not carefully managed, public access may affect farm quality assurance schemes and prejudice specialist breeding programmes. Sporting interests and game conservation may also be adversely affected by ill-informed access choices. Furthermore, the liability implications of injuries to members of the public continue to be of serious concern to land managers.

Notwithstanding these concerns and despite the range of issues that remain unresolved, land managers will play a pivotal role in the implementation of public access, and their views on it are vital to the success of the legislation. Often neglected when they are seen to run counter to the ethos of the new access rights, land managers' views reveal a wealth of experience that may help to anticipate likely problems and conflicts, and to form the basis of tailored interpretative materials for their amelioration. Research undertaken on behalf of SNH in 2000 (Land Use Consultants, 2000) explored the impact of the proposed new legislation on land managers in seven local authority pilot areas across Scotland. Focus group discussions were held in each of the areas and land managers were encouraged to comment on previous experiences of access and to express their views on the proposed legislative change. These discussions showed that many land managers were unaware of the legislation, and almost all knew nothing about the scope for additional support that is likely to accompany the new right of responsible access. There was also a perception amongst many of the land managers and owners that their concerns are neither fully understood nor respected by the wider public. Whilst land managers are in no way a homogeneous group, a number of shared experiences emerged during the course of the consultation. Consultees were primarily concerned with protecting their already fragile incomes, and were worried that increased access might further jeopardise the viability of their operations. At the same time, however, there was understanding and agreement that the majority of problems for land managers were generated by a small minority of users.

The consultation also shed light on some of the social differences in the way in which users behave. In urban fringe areas, problems of malicious and anti-social behaviour were clearly higher than in more remote rural areas, where conflict arose predominantly from a lack of understanding of the operational needs of land managers. However, land managers also recognised processes of counter-urbanisation and the changes that accompany such demographic shifts. Many newer 'rural' community members have no links with the farming communities within which they live (and vice versa) and are consequently viewed as being likely to cause problems on farmed or managed land, whether inadvertently or deliberately. Indeed, the feeling that people are unable to act responsibly where they have little understanding of the operational needs of land managers formed a common theme throughout the discussions for the pilot project. In relation to unintentional damage and interference, managers stressed the need to educate people about the seasonal sensitivities arising from operations and activities (e.g. gaming seasons, lambing, planting and harvesting) and the importance of not interfering in specific activities such as muirburn (the prescribed burning of moorland for management purposes) or stock management.

The promotion of user responsibilities is therefore likely to make a significant difference to the ways in which land managers respond to new access rights, and education was considered by land managers to play a central role in achieving this. There was also cautious recognition on the part of land managers that communities and other access users may be willing to learn more about their work and that, in turn, they may have a central role to play in facilitating this.

Recreational Users' Expectations of the Countryside

In addition to the concerns of land managers, it is important to reflect on the characteristics of recreational countryside users to provide a balanced view of the context within which the concept of responsible access is set. Motivations to holiday, seek recreation and pursue leisure in the countryside can vary widely from the need to escape urban pressures (*push* motivators) to a need for relaxation or self-fulfilment in a new setting (*pull* motivators) (Ryan, 1991: 31). Indeed, a whole range of motivational issues needs to be considered to provide a comprehensive picture of tourists' motivations (Sharpley, 1999: 131) and the general assumption (adopted here) is that motivations to visit (the countryside) will give rise to different needs and thus different types of behaviour to satisfy them. Whilst Crompton (1979) found that the push motive of 'a break from routine' did not necessarily require a change in lifestyle, Ryan (1991: 27) recognises the potential for a holiday to provide an opportunity for freedom from the normal constraints of home. Goffman (cited in Rojek, 1993: 165) puts forward the theory of 'action places' where individuals engage in forms of 'licensed revelry', and many current leisure forms fit Goffman's descriptions of such places where 'the rules of everyday life are relaxed and the boundaries of social behavioural are rolled back offer[ing] the experience of momentary escape from the encumbrances and pressures of everyday life' (Goffman in Rojek, 1993: 165). Urry (1995: 188) also believes that tourism is a conscious state in which conventional calculations of safety and risk are disrupted. Each of these 'states', therefore, offers potential for resistance to attempts to restrict personal choice and action within life's few spaces for inner direction.

Additionally, it must be recognised that visitors to the countryside are greater in number than in the past (EuroBarometer, 1998; Countryside Agency, 2000; VisitScotland, 2001) and that they use it differently as a visitor resource. Many 'traditional' pursuits such as walking and nature tourism for example are shadowed (and over-shadowed?) by more contemporary activities such as off-road driving, munro-bagging and survival exercises (see Butler, 1998; Roberts and Hall, 2001). The latter, in representing an imposition of urban values on rural space, are much less related to the character of their setting than 'traditional' pursuits, create different relationships with the rural context and demand different management techniques as a result. The existence, from the outset, of a draft code of conduct, suggests that this is understood. However, both the new forms of rural tourism and recreation, and contemporary meanings attached to such leisure by participants clearly present behavioural codes with considerable challenge, and it is therefore reasonable to assume that reconciling land manager concerns with recreational user aspirations is likely to be a complex process.

Codes of Conduct in Tourism

Evidence of the use of 'persuasive' communications in tourism to effect behavioural change can be found in the growing number of codes of conduct (Mason and Mowforth, 1996). They have been developed in response to the widespread and growing problems of tourism and recreation development in vulnerable environments. As noted from the land managers' perspective above, many of these traditional attempts at communication in order to influence tourists' behaviour have been seen to fail, evidence being manifest in continued land erosion, livestock damage, littering and inexorable pressures on the privacy of local communities. Perhaps in attempts to be understood by all, many of the codes favour a direct style with informative content presented as instructions to visitors (Mason and Mowforth, 1996), i.e. they are based on procedural information (Schultz, 2000) that is very unlikely to lead to the desired behavioural change. Additionally, a 'hectoring' approach may be counter productive, appearing offensive, prescriptive or patronising, and is unlikely to have a favourable impact in a leisure context.

Behavioural codes are important forms of social contract and the success of any such contract depends upon the social conditions within which it is to work. The liberalist tenet that individualism was the most desirable basis for behaviour in society was built on the underlying assumption that a social system based on any other view was naively utopian and doomed to failure (Kingdom, 1992: 6). As such, any sense of community was a means to an end rather than an end in itself. In reasserting a communal ethic, the enlightenment philosophers recognised principles of fraternity and equality in addition to liberty. One such view, that of Rousseau, believed in social conscience and altruism as well as self-interest. According to Rousseau, personal existence is relative rather than absolutist and self-interest is therefore seen in terms of the common good. People thus respect collective needs and express responsibility to them.

The idea of personal responsibility is critical to much reflexive political discourse. The re-emergence of liberalism in the years of the Thatcher administration was expressed in economic restructuring that embraced monetarism and the rolling back of the role of the state (Hutton, 1996: 90). An outcome of the re-emergence of a market philosophy in the public domain and the resulting shift of responsibility from the state to the individual, has been to require people to take more responsibility for their actions (Smart, 1999: 89). One of the long-term effects of increasing personal responsibility is an enhancement of self-interest and a concomitant reduction in civism. Therefore, a reduction in state welfarism is reflected in the outlook of individuals who are less inclined to behave altruistically. 'As an intrinsic concern for others has become more marginal, so the expectation of assuming/exercising moral responsibility has diminished' (Smart, 1999: 89). Released from the responsibilities of everyday life, whether seeking rest and renewal, excitement, pleasure or socialising, countryside visitors as tourists and recreationalists may be reluctant to conform to the requirements of behavioural codes redolent more of the constraints of everyday regulated existence than of purchased holiday freedoms. The broader societal context in which the land reform legislation has been conceived may therefore be seen to militate against ease of implementation because of the potential difficulties of inculcating the senses of community and altruism required to support the concept of 'responsible' access.

Codes have, however, been found to have a significant role to play in influencing behaviour (see Foley *et al.*, 2001: 113) although that role may not be easy to identify. If they are to become more effective, however, codes must learn to communicate by means other than basic exhortation, and the limits of the new Scottish Outdoor Access Code will be rigorously tested by contemporary social, leisure, recreation and tourism contexts.

Influencing Visitor Behaviour

The varying perspectives on countryside access underline the complexities arising from the different values and expectations of various user groups, as well as the difficulties of addressing the concerns of each perspective in its own right. Not only are land managers and users potentially at odds as a result of the ways in which they regard the countryside as a resource, but also, by their hedonic motivations and the leisure context of their activities, recreational users may be difficult to direct towards responsible behaviour where it conflicts with desired behaviour. As a result, it is important to give further consideration to the interpretative strategies that might be used to achieve the aims of the legislation through the Scottish Outdoor Access Code.

There is little explicit evidence of the existence of 'green' tourists who, aware of the potential impacts of their activities on social and physical environments, are prepared to modify their behaviour to reduce undesirable effects (Roberts and Hall, 2001: 142; Swarbrooke and Horner, 1999: 203). People's perceptions of the countryside vary, both in terms of its intrinsic value and as a resource for their enjoyment and it would therefore be wrong to assume that all tourist behaviours can be influenced in the interests of the environment. Like land managers, tourists are not a homogeneous, single minded group; the tourist condition not being one single state (Mason and Mowforth, 1996). It is therefore likely that a considerable education and interpretation programme will have to accompany the implementation of the new legislation in Scotland to ensure that all access is exercised responsibly.

According to the theory of reasoned action (Fishbein and Manfredo, 1992), many behaviours are under volitional control and therefore determined by the intention to perform the behaviour in question. This theory is primarily concerned with identifying the factors underlying the formation and change of intentions. It assumes that people are capable of processing information systematically and has three main hypotheses:

1   that behaviour is linked to intentions;
2   that intentions result from varying degrees of attitudes and subjective norms; and
3   that attitude and subjective norms create behavioural and normative beliefs.

A person's intention to perform a certain type of behaviour (e.g. walkers making a lengthy diversion rather than taking a direct route across farmland during lambing time) is a direct function of their belief that performing the behaviour will lead to pre-determined outcomes (that livestock will not be placed under stress as a result of their presence) and of the person's evaluation of these outcomes. The more a person believes that a certain type of behaviour will lead to positive outcomes (lack of livestock stress) or prevent negative ones (farmer conflict and associated guilt), the more favourable will be the attitude towards it. Generally speaking, a person will intend to perform a certain type of behaviour when they have a positive attitude towards it. If a person believes strongly that it is her/his responsibility to care for livestock on land being visited, s/he will do so despite the fact that others may not.

Additionally, behaviour is linked to a person's subjective norm whereby it becomes a function of his/her normative beliefs, in turn influenced by what significant others believe to be right, and the motivation to comply with such a norm. A less strongly held personal belief in the importance of livestock care, therefore, may make a decision more susceptible to influence from the subjective norm – the fact that access is not denied and others are using the footpath mean it is not unacceptable to do it. Where perceived volitional control is low and people feel their sole contribution cannot make a difference, they may change attitude but not necessarily behaviour (i.e. they may believe that it is undesirable not to use the footpath but join others crossing the field).

Following the theory of reasoned action, therefore, one is more likely to be successful in effecting a change in a given intention if one alters both individual attitudes and more widely held normative beliefs that correspond to it (Fishbein and Manfredo, 1992: 35; Smart, 1999).

The Roles of Interpretative Materials

Thus far, this chapter has noted the competing interests that are accentuated by the new access legislation in Scotland and the potential difficulties of effecting behavioural change in a recreational setting. Its focus now moves to the promotion of responsible access through the Scottish Outdoor Access Code and considers the ways in which these contextual considerations can contribute to its successful implementation.

'Interpretation is the process by which information is used as the basis of communication that aims to reveal meanings, challenge thought, and inspire imagination' (see Tilden, 1957). Good interpretation is illustrated by skilful communication of themes that can be linked to people's knowledge and personal experiences (Ham, 1992). Within countryside recreation management, interpretation is used as a visitor management tool to enhance visitor enjoyment and minimise site impacts. That is, it is used to persuade visitors to behave in a manner that is appropriate to the requirements of the site. Within the tourism context research into interpretation's potential as persuasive communication is not extensive, but that which has been done relates almost exclusively to tourism and the natural environment. Among countryside and heritage managers and providers, a growing interest in communicating with rural visitors has come either from a need to attract visitor spend or from a belief that visitors to the countryside need help to understand and care about it (Barrow, 1993: 271). Similarly, a number of studies has been conducted, some in response to the perceived needs of eco and nature-based tourists (Travelweek Australia in Orams, 1996), and others in the interests of environmental concerns (Moscardo, 1996; Orams, 1996, 1997; Stewart *et al.*, 1998).

Fundamentally, promotion of the Scottish Outdoor Access Code needs to persuade countryside visitors to care for the rural resource as they use it. The issue of care for the countryside using interpretative materials is approached by Stewart *et al.* (1998) in relation to *sense of place,* their question being 'does interpretation have the capacity to take visitors one step beyond wherever they happen to be when they arrive at a site so that empathy or "care" is developed towards the conservation of that place and sequentially to other places throughout the world.'

The study identified four main categories of interpretation users, illustrating a further layer of complexity in the behavioural change process. These are shown in Table 5.1.

Table 5.1 A typology of interpretation users

Seekers	visitors who actively seek out sources of information and interpretation
Stumblers	those who come across interpretative material by accident
Shadowers	those chaperoned by others through the interpretation that is offered (either formally or informally)
Shunners	visitors who either show no interest in interpretation or deliberately avoid it

Source: Stewart *et al.* (1998)

Stewart *et a.l*'s (1998) study points out that although interpretation may result in enhancement and enrichment of people's experiences of place, we know little about how this is actually achieved. We do know, however, that volitional behaviour does not change without a preceding change in attitude (Fishbein and Manfredo, 1992). Attitudes may be influenced by new information, appeals to emotion, and/or peer pressure resulting in the formation of new attitudes and subjective norms, and subsequent behaviour modification (Engel *et al.*, 1995: 387). But behaviour change is more complex still because attitude change does not result in behaviour modification by default. Research into recycling of domestic waste in the USA (Werner and Makela, 1999; Schultz, 2000) shows that recycling habits may be positively influenced by knowledge. The distribution of educational materials on recycling habits was found to lead to an increase in knowledge which, in turn, led to desired behaviour. However, the focus on procedural and impact knowledge (telling people what to do and explaining the effects of non-compliance) resulted in only small changes in behaviour that were short lived (Schultz, 2000). Although lack of information was found to be a barrier to recycling, information provision, whilst removing the barrier, was not sufficient to effect the desired behavioural change. According to Schultz (2000) this is because codes based on procedural and impact knowledge fail to accommodate people's motives. A focus on the inculcation of 'normative knowledge' on the other hand (as outlined) can develop an appreciation of 'injunctive' social norms – a set of beliefs about the 'right' things to do in given circumstances. This broader and more fundamental approach takes account of people's reasons for being in the countryside and their appreciation of its multifunctional use.

The development of a *sense of place* underpins the tourism management process at destination level as explained by Dagnall and Atkinson (1998), who believe that recreational use of land and water should be founded on care so that those who enjoy the outdoors are encouraged to respect its beauty, its wildlife, its operational needs and privacy of those who live or earn their livings there. Although Dagnall and Atkinson's (1998) paper recommends the establishment of voluntary codes of conduct for site management, the use of interpretative methods is implicit in this process.

Moscardo (1996) also believes that effective interpretation can alter behaviour, either directly through information or indirectly by fostering visitor appreciation of a site (using a normative approach). She reports on interpretation and built heritage management and believes that interpretation induces a state of mindfulness (Langer, 1989b; Langer *et al.*, 1989) essential to learning.

> Mindful visitors will understand the consequences of their actions and be able to behave in ways that lessen their impacts on a site. Mindful visitors will also have a greater appreciation and understanding of a site, and such understanding can provide support for both changing their behaviours on site and for the conservation of the site (Moscardo, 1996).

There is a positive link between visitor enjoyment and visitor learning (defined as mindfulness) (Moscardo and Pearce, 1986). Importantly, the state of mindfulness can, Moscardo (1996) claims, be induced by effective interpretation. As a counterpoint, reservations have been expressed by Orams (1996) who recognises the optimism involved in hailing interpretation as an effective tool in the management of nature-based tourism. Furthermore, research reviewed by Langer (1989a) indicates that mindlessness may be an extremely pervasive condition lending further support to the view that self-interest may occlude the welfare of others. A further study of interpretative material by Roggenbuck and Passineau (1986) evaluated the effectiveness of interpretation in determining positive attitudes and behavioural intentions, reductions in depreciative behaviour and stimulation of increased interest in resources in the USA. It showed varying results across groups (illustrating the difficulties of measuring the contribution of interpretation to behaviour change) but, generally, found a positive role for interpretative materials in increasing knowledge gain and developing positive attitudes and behavioural intentions.

Visitors' motivations also have a role to play in the extent of involvement in the consumption of the recreational product. It may be reasonable to assume that a visitor motivated more by a *pull* factor to a specific place than a *push* factor merely to escape routine will feel more of a sense of involvement with the specific destination (Roberts and Rognvaldsson, 2001: 96). If found to be effective, it should be the role of interpretation to develop that sense of place in all visitors in order to help them to understand the location and develop a sense of care towards it. The assumption is that the greater the extent of involvement in the process of tourism and the experience in relation to location, the more likely it is that a countryside visitor will be open to attitude and behaviour change when exposed to new information about it.

More recently, research in the field of interpretation and behaviour change has emphasised the importance of understanding visitors' needs and motivations, prior knowledge, attitudes and beliefs in relation to a site before attempting interpretive design (Ballantyne *et al.*, 1998). In other words, if interpretative materials are to be effective, they must contain communication strategies that create links between visitor and theme, and enable people to establish personally meaningful connections with the interpretative experience. This requires that material should be targeted at different types of visitor such as those categorised by Stewart *et al.* (1998).

It may also be the case that the interpretation process presents a means of reducing polarisation and overcoming conflict between land managers and users of access by the involvement of the former in the development of appropriate interpretative materials and user advice. As noted previously, land managers are not only well placed to report on problems associated with irresponsible access, but also to advise on the potential solutions. This aspect of the debate extends beyond the potential to reconcile competing interests in the countryside by connecting with much wider discourses on the urban/rural divide. As a result of ongoing problems being faced by the farming industry throughout Europe in recent years, many land managers are now recognising that there is a need to reconnect with broader society as agriculture continues to change. In securing a more positive approach to access provision, management and interpretation may contribute towards raising awareness of the role of land managers in environmental stewardship.

Conclusion

This chapter has considered the context of competing interests in, and perceptions of, responsible access within which the new legislation on access and the Scottish Outdoor Access Code will be implemented. It has reflected on the concerns of land managers as 'hosts' to recreational users and the consequent importance of promoting user responsibility if these fears are to be allayed. At the same time, however, the discussion aimed to emphasise the importance of considering the nature and social context of rural visiting as well as visitors' motives in tourism and recreation consumption. Emphasis was placed on the need to develop informational and/or interpretative materials that focus not only on individual attitude, but also on 'normative knowledge' i.e. the development of positive attitudes towards countryside conservation in general and the resulting need to behave 'appropriately'. For example, in order to encourage sensitive use of footpaths, effective interpretative material would deliver two messages – one to influence attitudinal considerations (the way an individual feels about being personally responsible for livestock stress) and another to influence social norms (the way people should feel about visitors' responsibilities generally).

Motivation, mindfulness, knowledge, understanding and attitudes are all inter-related influences on visitor behaviour and must be accommodated in the design of interpretative programmes intended to effect behavioural change. Land managers have the potential to become key actors in facilitating this process, thereby ameliorating some of the problems that they anticipate arising from wider access rights. Given increasing and changing use of the countryside for recreational purposes, the new Scottish Outdoor Access Code, which essentially replaces the longstanding Countryside Code, represents only part of what will be needed to explain and motivate responsible use of the countryside.

Note

1   Concurrent access legislation in England placed similar demands on the countryside but on land with very different ownership patterns and for more limited access: the proposed access rights in England and Wales were to mountain, moor, heath, down and registered common land only. As a result, legislation in England, although administratively complex, was rather less contentious.

References

Ballantyne, R., Packer, J. and Beckman, E. (1998), 'Targeted Interpretation: Exploring Relationships Among Visitors' Motivations, Activities, Attitudes, Information Needs and Preferences', *Journal of Tourism Studies*, vol. 9, pp. 14-25.

Barrow, G. (1993), 'Environmental Interpretation and Conservation in Britain', in F.B. Goldsmith and A. Warren (eds), *Conservation in Progress*, John Wiley & Sons, Chichester.

Butler, R. (1998), 'Rural Recreation and Tourism', in B. Ilbery (ed.), *The Geography of Rural Change*, Addison Wesley Longman, Harlow, pp. 211-32.

Countryside Agency (2000), *English Countryside Day Visits*, Countryside Agency, Cheltenham.

Crompton, J.L. (1979), 'Motivations for Pleasure Vacation', *Annals of Tourism Research*, vol. 6, pp. 408-44.

Dagnall, P. and Atkinson, C. (1998), *Towards Responsible Use: Influencing Recreational Behaviour in the Countryside*, an advisory note prepared for Scottish Natural Heritage by CEI Associates (unpublished).

Engel, J., Blackwell, R. and Miniard, P. (1995), *Consumer Behaviour*, Dryden Press, Fort Worth, 8th edition.

EuroBarometer (1998), *Facts and Figures on the Europeans' Holiday*, EuroBarometer for DG XXIII, Brussels.

Fishbein, M. and Manfredo, M.L. (1992), 'A Theory of Behaviour Change', in M.L. Manfredo (ed.), *Influencing Human Behaviour: Theory and Applications in Recreation, Tourism and Natural Resources Management*, Sagamore Publishing, Sagamore, pp. 29-48.

Foley, M., Frew, M. and McGillivray, D. (2001), 'Influencing Recreation Behaviour to Reduce Impacts', in M. Usher (ed.), *Enjoyment and Understanding of the Natural Heritage*, The Stationery Office, London, pp. 105-19.

Ham, S.H. (1992), *Environmental Interpretation*, North American Press, Colorado.

Hutton, W. (1996), *The State We're In*, Vintage Press, London.

Kingdom, J. (1992), *No Such Thing as Society? Individualism and Society*, Open University Press, Buckingham.

Land Use Consultants (2000), Feasibility Study: Local Authority Pilot Project, Land Use Consultants, Glasgow, unpublished.

Langer, E. (1989a), *Mindfulness*, Addison Wesley, Reading, Massachusetts.

Langer, E. (1989b), 'Minding Matters', in L. Berkowitz (ed.), *Advances in Experimental Social Psychology*, vol. 22, pp. 137-74.

Langer, E.M., Hatem, J., Joss, J. and Howell, M. (1989), 'Conditional Teaching and Mindful Learning: The Role of Uncertainty in Education', *Creativity Research Journal*, vol. 2, pp. 139-50.

Mason, P. and Mowforth, M. (1996), 'Codes of Conduct in Tourism', *Progress in Tourism and Hospitality Research*, vol. 2, pp. 151-67.

Moscardo, G. (1996), 'Mindful Visitors, Heritage and Tourism', *Annals of Tourism Research*, vol. 23, pp. 376-97.

Moscardo, G. and Pearce, P.L. (1986), 'Visitor Centres and Environmental Interpretation: An Exploration of the Relationships Among Visitor Enjoyment, Understanding and Mindfulness', *Journal of Environmental Psychology*, vol. 6, pp. 89-108.

Orams, M.B. (1996), 'Using Interpretation to Manage Nature-based Tourism', *Journal of Sustainable Tourism*, vol. 4, pp. 81-94.

Orams, M.B. (1997), 'The Effectiveness of Environmental Education: Can We Turn Tourists into Greenies?', *Progress in Tourism and Hospitality Research*, vol. 3, pp. 295-306.

Roberts, L. and Hall, D. (2001), *Rural Tourism and Recreation: Principles to Practice*, CAB International, Wallingford.

Roberts, L. and Rognvaldsson, G. (2001), 'The Roles of Interpretation in Facilitating Access to and in the Countryside', in L. Roberts and D. Hall, *Rural Tourism and Recreation: Principles to Practice*, CAB International, Wallingford, pp. 92-7.

Roggenbuck, J.W. and Passineau J. (1986), 'Use of the Field Experiment to Assess the Effectiveness of Interpretation', *Proceedings of the South Eastern Recreation Research Conference, Athens*, pp. 65-86.

Rojek, C. (1993), *Ways of Escape: Modern Transformations in Leisure and Travel*, Macmillan Press, London.

Ryan, C. (1991), *Recreational Tourism: A Social Science Perspective*, Routledge, London.

Schultz, P.W. (2000), *Knowledge, Education and Household Recycling, Examining the Knowledge-deficit Model of Behaviour Change*. Paper presented at a meeting of the National Academy of Sciences Committee on the Human Dimensions of Global Change, 29 November, Washington DC.

Sharpley, R. (1999), *Tourism, Tourists and Society*, ELM Publications, Huntingdon, 2nd edition.

Shoard, M. (2000), 'Access to the Countryside', Paper presented at *The Town and Country Planning Summer School*, St. Andrews University, Scotland, 6-13 September.

Simpson, F. (2001), 'Access and Land Management', in L. Roberts and D. Hall, *Rural Tourism and Recreation: Principles to Practice*, CAB International, Wallingford, pp. 86-8.

Smart, B. (1999), *Facing Modernity. Ambivalence, Reflexivity, Morality*, Sage, London.

Stewart, E.J., Hayward, B.M. and Devlin, P.J. (1998), 'The "Place" of Interpretation: A New Approach to the Evaluation of Interpretation,' *Tourism Management*, vol. 19, pp. 257-66.

Swarbrooke, J. and Horner, S. (1999), *Consumer Behaviour in Tourism*, Butterworth Heinemann, Oxford.

Tilden, F. (1957), *Interpreting our Heritage*, University of North Carolina Press, North Carolina, 3rd edition.

Urry, J. (1995), *Consuming Places*, Routledge, London.

VisitScotland (2001), *Tourism in Scotland 2000*, Scottish Tourist Board, Edinburgh.

Werner, C. and Makela, E. (1999), 'Motivations and Behaviours that Support Recycling,' *Journal of Environmental Psychology*, vol. 18, pp. 373-86.

Chapter 6

The Host-Guest Relationship and its Implications in Rural Tourism

Hazel Tucker

Introduction

This chapter addresses the host-guest relationship developed in small accommodation businesses in the rural tourism context. Drawing on 'exchange theory' (Mauss, 1967; Wood, 1994) and findings from research conducted in rural areas of New Zealand's South Island, the discussion highlights the importance, the implications, and also the fragility of the host-guest relationship in view of the ongoing sustainable development of rural tourism. Whilst an increasing body of tourism research has considered the economic issues surrounding the role of small businesses in rural development, relatively little attention has been paid to the host-guest relationship in this tourism sector, even though it is this relationship which comprises the key 'experience' being bought and sold, particularly in the case of small hosted accommodation enterprises. Research to date that has focused on the social dynamics of tourism hospitality in the Bed and Breakfast-style accommodation sector has been useful in drawing out many of the issues and tensions inherent in the notion of *commercialised* hospitality. These include the question 'Can a commercial host be hospitable?' (Telfer, 2000: 42), and the tension arising from the necessity of monetary exchange in the home-based tourist accommodation sector leading to the exchange and experience of 'genuine' hospitality being diminished (Heal, 1990; Lashley, 2000; Pearce, 1990).

The intention in this chapter, however, is to explore the implications that stem, in particular, from the intense social exchange that takes place in this setting, thereby making the host-guest relationship a focal point in addressing rural tourism. What, for example, are the implications of tourist visitors to rural areas becoming 'guests' in relation to their rural 'hosts', in the real rather than the tourism industry jargon senses of these two words? Furthermore, how does the imposition of host and guest identities affect tourist experience and behaviour in rural locales in view of the reciprocity, obligation and control that are an inevitable part of the social exchange in the host-guest relationship? The following discussion explores these issues drawing upon findings from research undertaken in the Central Otago and Marlborough regions of the South Island of New Zealand. This research is based principally on participant observation and in-depth interviews conducted with Bed and Breakfast and home-stay hosts and guests, as well as motel guests in the studied regions. It should be noted here that whilst there are some important differences to be drawn between 'Bed and Breakfasts' and 'home-stays' in this region, they have been pooled together for the purposes of this chapter in which the crucial distinction is between the expectations regarding setting, service and 'guest'-'host' interaction in these small home-based establishments compared with the larger hotel/motel-style accommodation (see Darke and Gurney, 2000; Ombler, 1997).

Context

Bed and Breakfast (B&Bs), home-stays and farm-stays represent a rapidly growing sector in New Zealand's tourism industry. Throughout the 1990s B&B-style accommodation grew rapidly, and now there are estimated to be at least 640 B&B-style businesses throughout the country (Hall and Rusher, 2003), although this figure is difficult to ascertain exactly because of the casual propensity of businesses in this sector. In New Zealand, B&Bs tend to be as or more expensive than much motel and hotel accommodation, indicating that both the hosts and guests see this type of accommodation as the buying and selling of an 'experience' (Pine and Gilmore, 1999; Roberts and Hall, 2001) rather than 'just a bed'. Indeed, the majority of the visitors choosing B&B-style accommodation in rural New Zealand are international FITs (free independent travellers) who want to experience the 'real' New Zealand by getting off the beaten track, meeting New Zealanders, and experiencing aspects of New Zealand apart from the usual tourist icons (Collier, 1997).

Most B&B guests come from the USA, UK and other parts of northern Europe. A limited number of Japanese and other east Asian nationalities stay in this type of accommodation, plus a very small, though reportedly growing, number of New Zealanders. In interviews, some general reasons guests reported for staying in B&Bs and home-stays included the tendency for this type of accommodation to be peaceful, to be in beautiful settings, to be 'homely', unique and unlike the 'bland' barn-like feeling experienced when staying in motel accommodation. The overriding reason for this choice, however, was 'to have a relationship with local people', 'to have the opportunity to talk with them', 'to get to know the lifestyle of New Zealanders' and 'to learn about their culture'. In other words, the hosts themselves are a key attraction in this accommodation sector, and home-hosted accommodation is perceived by tourists to provide them with an apparently 'back-stage' experience of rural New Zealand life.

As well as the visitors' reasons for choosing to stay in B&B-style accommodation, the motivations for people living in rural areas to operate a B&B or home-stay business in their home are important to consider since the motivations of both parties are centred in their host-guest interactions (Lynch and MacWhannell, 2000; Pearce, 1990; Stringer, 1981). The motivations that hosts reported in interviews seem to match well with those of their guests: social reasons are primary whilst economic motivations are largely played down (similar findings were obtained by Pearce (1990) in relation to New Zealand farm tourism, though Pearce also rightly noted that these statements by hosts should be interpreted cautiously because of the social desirability factor concerning responses to questions on financial matters). Whilst many B&B and home-stay hosts did cite income as a motivating factor, especially as many were using that income to help pay towards and justify the large size of their home, most of them also strongly asserted social motivations for taking in guests. Opening up their home to tourist guests allows rural people the chance for interaction with different cultures and ideas that they might not otherwise have as part of their daily life in a rural community. Many of the hosts interviewed stated that they enjoyed 'having the company' and 'meeting different people and sharing ideas'. Furthermore, over half of the hosts who participated in the research were retired couples whose grown-up children had left home leaving space (both physical and social) that might be filled by paying guests.

Importantly, many hosts saw themselves as being the 'right' people to run B&B or home-stay accommodation, as they considered themselves to be 'people people' and thus suitable to host guests. Most of the hosts suggested in interviews that their role as hosts placed them in the position of being representatives of the New Zealand people (and, more specifically, 'ordinary' New Zealand people), and they felt particularly suited to the job if they felt they held a certain aspect of local knowledge, such as history or wildlife, where to go fishing or home-preserve making. These areas of knowledge allow hosts to perform a rural New Zealand identity to their guests, portraying New Zealand as a rural 'pastoral paradise' (Conrich and Woods, 2000). As said above, the hosts in this setting are clearly a focus of the tourists' attraction, in their representing of 'rurality' (Roberts and Hall, 2001) and, particularly in the case of New Zealand, a 'pastoral' nation. Issues pertaining to the hosts' identity thus extend the boundaries of their notions of hospitality beyond the home and outwards to the local area and even to the nation, and, as will be explained later in this chapter, this is important to consider when exploring issues of reciprocity, obligation and control that emerge from the host-guest relationship in this rural tourism context.

Reciprocity, Obligation and Control

As Selwyn (2000: 19) points out, 'The basic function of hospitality is to establish a relationship or to promote an already established relationship'. That is, the giving and receiving of hospitality (food and drink, accommodation and entertainment) engages principles of reciprocity between hosts and guests, and thus a complex set of interactional rules involving shared values and trust (Burgess, 1982; Mauss, 1967; Selwyn, 2000; Wood, 1994). Hospitality is hence a transformation process, wherein the key transformation taking place is that from a set of strangers into friends. Indeed, most of the hosts interviewed described this transformation process when discussing their guests. For example, one home-stay host said in conversation, 'it gives me a great buzz that perfect strangers can come on in and I can see within an hour they feel quite relaxed'. Many hosts stated that they saw their guests as 'guests' when they first arrived but as 'friends' when they departed.

As stated in the previous section, meeting people and social interaction is the main reason both hosts and guests give for wishing to enter into this relationship in the B&B and home-stay setting. It is, however, a highly complex relationship to enter into, as the interrelationship between hosts and guests is governed by an extensive set of social rules and obligations. Moreover, since the context of this type of tourist hospitality provision is in hosts' own homes, the rules are also strongly mediated by the meanings surrounding the home (Darke and Gurney, 2000; Lynch and MacWhannell, 2000). By letting 'strangers' stay in their home, B&B hosts are taking a variety of risks and must therefore take certain measures to ensure that their guests will play by the rules, let alone understand the rules in the first place.

It is arguable that only a particular 'type' of tourist will opt to stay in B&B-type accommodation, and this is the initial measure of social control that the operation of B&B-type accommodation in rural areas puts in place. Just as Darke and Gurney (2000) state that commercial home-based hospitality in Britain is a middle class phenomenon, so too might B&B hosting and staying in New Zealand be considered specific to a particular level of affluence and education. Indeed, Tourism New Zealand claims that today's international visitors to New Zealand are environmentally, culturally and socially savvy, being well-travelled and having a global mindset (Hickton, 2003). The particular tourists who choose to stay in B&B accommodation would therefore appear to be New Zealand's ideal type of visitor. When discussing their guests, hosts state that they generally get 'nice people' staying, people who 'come to learn', and who are 'interested'. Pearce (1990) has also stated, more specifically in relation to farm-stay tourism in New Zealand, that there is a certain pressure on guests to be entertaining and knowledgeable about their home country. Home-stay guests, in other words, must be the type of people who are practised at dinner party conversation, for they will hopefully understand the rules of the host-guest relationship.

Both the hosts and the guests, then, take on a mutual responsibility to ensure that their interaction runs as smoothly as possible for the duration of the visit, so that the transformation process, from stranger to friend, is carried out effectively. For this reason, particularly at the beginning of each guest's stay, hosts will often tell many anecdotes and jokes in order to socialise the listener and to reaffirm that the hosts and guests share the same understanding of social norms (Darke and Gurney, 2000). Additionally, throughout the stay if guests are suspected of deviating from the usual rules of play, hosts may continue to tell stories of previous guests' inappropriate behaviour so as to serve as a warning in order to draw their guests' behaviour back into line. This was illustrated to the researchers during our participant observation in relation to appropriate conduct when returning to the home-stay after dining out in the evening. After staying out perhaps slightly later than the established 'norm' on our first evening's stay, we were told over breakfast the following morning repeated stories about other guests who had stayed out late and come in drunk and making a lot of noise. Whilst we were assured by our hosts that we ourselves had not stepped over the line, these stories let us know clearly where that line was. We heard such stories regularly during our stays in various B&Bs concerning matters ranging from guests using up the hot water, to helping themselves in the kitchen, or complaining about the breakfast, to cutting their toe-nails in the living room.

Interestingly, when asked in interviews about any problematic or awkward situations that might be recalled, very few breaches of the codes of conduct were reported by hosts or guests, which points to the fact that guests do tend to generally play by the appropriate rules with little serious deviation. Moreover, this indicates that throughout their stay guests are very much obliged to abide by their hosts' rules because of the principles of reciprocity which are engaged through the exchange of hospitality. As Heal (1990: 192) explains, the guest is obliged 'to accept the customary parameters of his hosts' establishment, functioning as a passive recipient of goods and services defined by the latter as part of his hospitality'.

The hosts therefore hold a position of significant control over their guests, and this control emerges specifically from the *social* exchange of hospitality and the relationship which hence arises between the giver and the receiver. The control afforded to hosts through their giving hospitality tends to start from the moment the guests arrive. When asked, for example, what happens upon guests' arrival, one set of hosts answered: 'We always offer them a hot drink or a glass of water or whatever, and a muffin or a cookie or something and ask if they would like to sit down and join us, and that to me is a very important time. To us it is a very important time. Because it is when you start to build up a relationship and find out what they want to do'.

The ties of obligation impressed upon the guests then continue throughout their stay. Guests are in the hosts' space and so as well as abiding by the general social rules which ensure the interaction will run smoothly, they are also expected to respect and submit to their hosts' way of doing things. As one host explained, 'they are aware that they are in our situation at the time so they are wanting to be like us, or accept the way we do things'. This idea was reported by both hosts and guests in interviews. Just as hosts acknowledge their guests' role in accepting the hosts' parameters, guests are fully, perhaps even intensely, aware of the relationship they have entered into by staying in a home-stay situation: 'In someone's home you feel more conscientious about tidying up after yourself, and you have to hang around longer – you can't up and leave like in a hotel'.

Hosting for the Rural Area

By being 'guests' in a home-stay, tourists thus enter into a relationship which rids them of the sense of tourist freedom and anonymity they would have when staying in a hotel or motel. Moreover, they become 'guests', and have to abide by their hosts' parameters, not only within their hosts' home, but also beyond the home and to the surrounding areas where the hosts feel that they are representative 'hosts'. One host stated this clearly by saying: 'We hope they leave feeling they've enjoyed our treatment of them, enjoyed our hospitality, which leaves them feeling good about Wanaka [the name of the town in which their home-stay was situated] and New Zealand'.

Indeed, most hosts extend their help and information provision at least to the local region, and many also informally guide their guests around the local area. As another host couple answered when asked about this matter: 'Well, we certainly like to promote the area and what's here, definitely. We quite often go for walks with them. We like to promote the area and help them with it...A lot of them might feel they don't know where they're going, what's going to happen, which is understandable, so I say "I'll come with you"'.

Guests actively seek this sort of 'local knowledge' from their hosts. Such knowledge and guidance, as well as the hosts' ability to 'personally' recommend a good local restaurant, was often cited by guests as part of their reason for staying in B&B-type accommodation, particularly in the more rural areas of the country.

Importantly, though, the personal exchange of this sort of information further locks B&B hosts and guests into a complex negotiation involving trust and obligation. Firstly, the guests' trust of the recommendations and information they receive is based on an awareness that their hosts will have to face the repercussions of their recommendation. A number of guests interviewed stated the idea that: 'You respect the likelihood of honesty of what they're telling you simply because they have to deal with you after it's all over'. Simultaneously, however, guests are obliged *not* to return and complain to their hosts that they were given bad advice. To keep relations smooth they are obliged, rather, to return and bestow the virtues of the recommended place, and to tell their hosts, over tea and muffins usually, about the wonderful time they've been having. Furthermore, the guests are obliged to actually go through with the recommendation provided by their hosts. When the information and advice given to guests is given through the idiom of hospitality, it would be inappropriate to ignore it and do something entirely different (Tucker, 2003). Perhaps it is for this reason that many of the hosts interviewed and observed were hesitant to suggest particular advice and recommendations in the local area surrounding their establishment. Rather they preferred to 'find out first what kind of things the guests are interested in, and then give them a few options'. This would avoid pinning their guests down to absolutes and thus save potential embarrassment to both host and guest.

It is thus the *potential* level of control B&B and home-stay hosts can have over their guests' behaviours and experiences in the wider rural area that is important to consider, this control begin created by the obligation the guests are under to do what their hosts suggest. In a two-pub town, for instance, where one particular pub owner was out of favour with local residents, it would be likely that all visitors staying in B&Bs in the town would only drink at the pub that was in favour. Similarly, hosts who had particular opinions regarding environmental issues in the area could be pretty persuasive in ensuring that their guests did not walk a path that they saw as being environmentally damaging. In rural areas, B&B and home-stay guests are locked into a relationship that goes beyond the physical walls of the hosts' actual house because the obligation in hospitality extends to the wider meaning of 'home', and 'home' for the hosts is also the village or town and the surrounding region. The host-guest relationship thus has important implications for rural tourism management and the continuation, or sustainability, of certain social, business and natural tourism environments.

Restricting Freedom and Independence

A final aspect of this sustainability, however, is the question of how the host-guest relationship and the balance of obligation and control it engenders is experienced by the guests – the visitors to rural areas who enter into this type of relationship. If, when tourists become 'guests', they relinquish significant control over their activities and behaviour to their hosts, how does this affect their experience of the rural area they are in? On one hand, this more intense social interaction and receiving of local knowledge through the idiom of hospitality is the reason that many 'free independent travellers' choose the B&B and home-stay experience. On the other hand, the host-guest relationship entered into presents an immediate paradox. For although these FITs are free and independent from the organised packaged tour, making their own choices regarding their itinerary, mode of transport and accommodation, they can easily become subsumed instead by their B&B hosts, and find the freedom they were aiming at undermined by their being 'guests', and therefore in a position of obligation, in relation to their hosts. The positive aspects of experiencing 'Kiwi hospitality' in a New Zealand home setting can very easily be out-weighed by a negative sense of restriction and obligation.

It is precisely the awareness of this restriction and the social ties inherent in the host-guest relationship which leads many travellers to choose to stay in motels. As Lashley (2000: 12) has noted, 'The commercial and market driven relationship which allows the customer a freedom of action that individuals would not dream of demanding in a domestic setting is one of the benefits claimed for the "hospitality industry".' When motel guests interviewed were asked why they chose this type of accommodation over B&Bs or home-stays, besides a general hesitancy about trying something 'new', there were in most of their replies the notions of privacy and independence: 'We're not B&B people – we like our independence, we don't like to front up to breakfast.'

The success of home-based accommodation businesses in attracting custom in the first place is thus limited because of the fear of the social ties inherent in the role of 'guest'. Furthermore, for those who do choose to experience home-based hospitality, the length of stay in any one establishment and place can be limited for the same reason. The longer a guest stays in a home-stay the stronger the social ties are likely to become with the hosts. This idea was expressed frequently by hosts when asked about the relative benefits of longer staying guests over one-nighters: 'If they stay a few days you get to know them better...you have a rapport with them and it's not like a "one-night-stand"'. Interestingly, many hosts interviewed stated that they preferred guests staying for only one night as any longer would be a restriction on their own freedom and privacy: 'I wouldn't like them to stay for two or three nights because I'd have to entertain them, I couldn't just watch TV'. Indeed, most B&B and home-stay guests in the rural regions studied do only stay for one night in each place, and by doing so, they enter into the host-guest relationship just enough to reap the positive aspects of the experience, and then they depart before the transformation process described above gets fully underway. Whilst they are keen that their identity be changed from 'tourist' to 'guest', they are usually keen to escape before they are fully transformed into 'friends'. A developing 'friendship' would muddy the commercial dimension of the relationship, giving rise to confusion regarding payment. Guests might wonder, for example, whether they will and should be charged for the dinner and wine they shared with their hosts on the second or third night of their stay. When the standing of the relationship becomes unclear, it is easy for both parties to offend and be offended. This also has implications regarding the issue of return visitation, as a social relationship developed within the B&B context may be too complex to contend with if guests are considering returning to the area at some later date. They may find it easier to not return to the area at all.

There is no doubt that commercial hospitality provision in the home is a highly delicate balancing act. This takes us back to the issue mentioned earlier of the paradox inherent in the notion of 'commercial hospitality' and the apparent tension in the experiences for guests who must pay a monetary fee for what is otherwise a social exchange. However, what this in-depth study of the host-guest relationship has highlighted is that rather than payment for hospitality necessarily being a problematic paradox in the rural tourists' and hosts' relationship and experience, the monetary exchange which runs parallel with the intense social exchange taking place is often an underlying comfort for those involved. By handing over the payment upon their departure from a home-stay establishment, guests are able to regain the freedom and independence they desire, and this may also help regarding potential future meetings if and when tourists make return visits. The payment marks something of a cleaning of the slate, so that 'commercial hospitality' may take place again between the two parties in the future.

Conclusion

The purpose of this chapter has been to draw attention to the host-guest relationship as an important factor in consideration of sustainable tourism in rural areas. The transformation of 'tourists' into 'guests' through the provision of hospitality can have significant implications regarding the behaviours and experiences of those guests, and also the continuing success of the area as a tourist-receiving environment. As it is only a certain 'type' of tourist who is willing to enter into the host-guest relationship by staying in B&B-type accommodation in the first place, then this type of accommodation, by default, selects the type of tourist visiting and staying in an area. Furthermore, once the host-guest relationship is established, the local 'hosts' can gain substantial control over visitors' behaviour and experience because of the principles of reciprocity and obligation inherent in this social form of exchange.

This argument is, of course, based on the host-guest relationship which, within the home-stay context, is in its most extreme form. It might be questionable, for example, as to just how extensive the host-guest repercussions can be in an area which only has a sprinkling of B&B and home-stay establishments. What is important, however, is to highlight the implications that the giving of hospitality can have generally in rural tourism. That hospitality can exist not only in the home-hosted accommodation sector of the tourism industry, but also in other tourism operations including adventure and eco-tour operations, guiding operations, and even visitor information centres. What this analysis of B&B and home-stay hosting has clarified is that even where there is an obvious presence of economic exchange, and the notion of an experience being 'purchased', this need not necessarily detract from the level of social exchange and the repercussions of that social exchange that can result.

Clearly, however, it is necessary for 'hosts' in rural areas to achieve the right balance regarding the extent to which they give hospitality to tourists. While visitors to rural areas often choose B&B or home-stay accommodation as a means to sharing the hosts' culture, hospitality and local knowledge, they may also be quick to experience unwanted feelings of restriction and obligation. Similarly, hosts themselves may experience a sense of invasion of privacy. The discussion in this chapter has been suggestive of some of the ways in which the correct balance might be achieved in rural hospitality provision, thereby maximising the success of the rural tourism experience for both hosts and guests. Firstly, whilst B&B and home-stay hosts, and indeed any other 'hosts', should be willing to offer help, service and information to guests, they should, at the same time, ensure that their guests are given plenty of 'space', privacy and independence without any guilt feelings attached for wanting that independence. Secondly, this can be achieved by hosts giving plenty of options for their guests both within the home setting and also when providing local information and recommendations. Finally, rather than trying to disguise the 'commercial' aspect of their intent in providing hospitality, perhaps hosts should recognise that commercial and social exchange can co-exist quite comfortably, and that, in fact, the commercial exchange is the let out clause for guests when the social intensity is such that they need to reclaim their freedom and independence from their rural hosts. In summary, the host-guest relationship is central to the experiences of both hosts and guests in this context and achieving the right balance in the relationship has significant implications for the future of tourism in rural areas.

References

Burgess, J. (1982), 'Perspectives on Gift Exchange and Hospitable Behaviour', *International Journal of Hospitality Management*, vol. 1, pp. 49-57.

Collier, A. (1997), *Principles of Tourism: A New Zealand Perspective*, Longman, Auckland.

Conrich, I. and Woods, D. (2000), *New Zealand, a Pastoral Paradise?*, Kakapo Books, Nottingham.

Darke, J. and Gurney, C. (2000), 'Putting up? Gender, Hospitality and Performance', in C. Lashley and A. Morrison (eds), *In Search of Hospitality: Theoretical Perspectives and Debates*, Butterworth-Heinemann, Oxford, pp. 77-97.

Hall, C.M. and Rusher, K. (2003), 'Risky Lifestyles? Entrepreneurial Characteristics of the New Zealand Bed and Breakfast Sector', in R. Thomas (ed.) *Small Firms in Tourism: International Perspectives*, Routledge, London and New York.

Heal, F. (1990), *Hospitality in Early Modern England*, Clarendon Press, Chicago.

Hickton, G. (2003), 'Tourism New Zealand Presentation to Dunedin Tourism Industry', Unpublished, Dunedin Centre, Dunedin.

Lashley, C. (2000), 'Towards a Theoretical Understanding', in C. Lashley and A. Morrison (eds), *In Search of Hospitality: Theoretical Perspectives and Debates*, Butterworth-Heinemann, Oxford, pp. 1-16.

Lynch, P. and MacWhannell, D. (2000), 'Home and Commercialised Hospitality', in C. Lashley and A. Morrison (eds), *In Search of Hospitality: Theoretical Perspectives and Debates*, Butterworth-Heinemann, Oxford, pp. 100-114.

Mauss, M. (1967), *The Gift*, Routledge, London.

Ombler, K. (1997), 'Motels vs Private Lodging ... Same Game ... Different Rules', *Hospitality*, vol. 33, pp. 6-10.

Pearce, P.L. (1990), 'Farm Tourism in New Zealand: A Social Situation Analysis', *Annals of Tourism Research*, vol. 17, pp. 337-52.

Pine, J. and Gilmore, J.H. (1999), *The Experience Economy: Work Is Theatre and Every Business a Stage*, Harvard Business School Press, Boston.

Roberts, L. and Hall, D. (2001), *Rural Tourism and Recreation: Principles to Practice*, CABI Publishing, Wallingford.

Selwyn, T. (2000), 'An Anthropology of Hospitality', in C. Lashley and A. Morrison (eds), *In Search of Hospitality: Theoretical Perspectives and Debates*, Butterworth-Heinemann, Oxford, pp. 18-36.

Stringer, P.F. (1981), 'Hosts and Guests: the Bed-and-Breakfast Phenomenon', *Annals of Tourism Research*, vol. 8, pp. 357-76.

Telfer, E. (2000), 'The Philosophy of Hospitableness', in C. Lashley and A. Morrison (eds), *In Search of Hospitality: Theoretical Perspectives and Debates*, Butterworth-Heinemann, Oxford, pp. 38-55.

Tucker, H. (2003), *Living With Tourism: Negotiating Identities in a Turkish Village*, Routledge, London.

Wood, R. (1994), 'Some Theoretical Perspectives on Hospitality', in A.V. Seaton, C.L. Jenkins, P.U.C. Dieke, M.M. Bennet, L.R. MacLellan and R. Smith (eds), *Tourism: The State of the Art*, John Wiley & Sons, Chichester.

Chapter 7

Animal Attractions, Welfare and the Rural Experience Economy

Derek Hall, Lesley Roberts, Françoise Wemelsfelder and Marianne Farish

This chapter briefly examines the nature and ethics of animal-based attractions within the tourism experience economy, and evaluates the role of interpretation in enhancing experience at animal-related tourism and recreation attractions. A rapidly increasing recreational demand for personal interaction with animals intersects with a growing awareness of our potential to disrupt the lives of animals and to negatively affect their welfare. There thus appears to be a potential conflict and contradiction in the employment of animals as 'attractions'. Yet objects of the tourist gaze are increasingly required to be managed and presented in a way which enhances and enriches the visitor experience.

As value-added mediation, interpretation has the potential to emphasise the experiential quality of directness of personal contact, and in the context of animal attractions can assist visitors to learn to interpret animals' actions and expressions. This chapter therefore argues that the role of interpretation in relation to animal-related attractions can be particularly significant in bringing substantial added value to the visitor experience in both explicit and ethical ways.

Introduction

The growth in popularity of 'nature-based' tourism, perhaps by 30 per cent per year (Young, 1998), is taking place alongside awareness of our impact on the environment in general and on wildlife in particular. Underlying the growing trend for wildlife-based tourism have been debates concerning the desirability of employing animals as objects of recreational 'edutainment', and on ways of minimising the undesirable impacts of such tourism (Orams, 1996a).

The tourism literature has often regarded 'wildlife' as a resource for sustaining business, aimed at providing the visitor with desired experiences from a range of animal-oriented activities. While this has embraced a concern for animal welfare in a far from consistent manner, it is placed largely within the context of the need to sustain the viability of the attraction, largely ignoring animals' longer-term responses to such 'non-consumptive' use (Curry et al., 2001). Further, until recently, the non-consumptive dimension of human relations with wildlife had received much less attention than that of wildlife as a source of food, clothing, trophies and other resources (Reynolds and Braithwaite, 2001).

Within this industry perspective, academic classifications of animal attractions have dealt only very generally with the need of visitors for personal interaction with animals. From an interactional and experiential perspective, such typologies have been unsatisfactory for at least two reasons. First, they provide little insight into the existing and potential experiential added-value of visitor-animal interactions. Second, they usually fail to embrace a perspective which suggests that experience for both parties can be enhanced through encouraging a greater understanding of animal behaviour and welfare.

Yet, the commodification of animals as part of mythic rural landscapes, as an integral element of the pastoral idyll, and as objects to anthropomorphise and adopt as children-friendly consumption goods, has fuelled the desire to experience different levels of interaction with animals. Notions of the experience economy – that memorable experience should be generated when providing a service or product – are derived from a realisation that visitors' experiences are drawn from those products and services that provide the stage and supporting backdrop, and that visitor experience can be presented as a 'distinctive economic offering' (Pine and Gilmore, 1998: 97). Of course, experiences have always played an important part in tourism and recreation markets, and many attraction managers, particularly those who have employed good and imaginative interpretation may find little that is original in this conceptualisation.

Drawing on two important dimensions of visitor orientation – participation, actively rather than passively engaging with an 'object', and connection, the degree to which a visitor is mentally absorbed or immersed – human interaction with animals has the potential to encapsulate all four realms of (human) experience as articulated by Pine and Gilmore (1998: 102) – entertainment, education, escapism and aestheticism. Such interaction offers an opportunity for a unique experience, and such an experience can provide the main objective of a visit or can be employed to transform positively recreational activities within a visitor attraction. While such apparent commodification may be distasteful for those for whom authenticity of experience is paramount, suggesting an 'aura of superficiality' (MacCannell, 1999: 98), it does have the power to alter – and enhance – the meanings attached to animals and their role in both 'wild' and 'captive' circumstances.

As Pine and Gilmore (1998: 104) put it: 'The more senses an experience engages, the more effective and memorable it can be'.   Indeed, this is a basic principle of good educational practice. And it should therefore follow that the range of senses drawn upon in the experience of 'appropriate' interaction with animals can transform an attraction visit. Employing the medium of 'appropriate' interpretation has a strong capacity to enhance the (human) experience of interacting with animals through permitting visitors to be more aware of an animal's behavioural and welfare needs, and acting accordingly, and thereby being closer to the experience of 'communication' with a sentient member of another species (e.g. see Wemelsfelder, 1997, 1999, 2001).

Animal-human Welfare in Tourism and Recreation

Although animal welfare and animals rights issues are important in the social sciences, there have been limited attempts to place them within a theoretical framework in relation to tourism and recreation. Exceptions include aspects of the addressing of sustainability questions in relation to marine life viewing, the establishment and maintenance of game parks and wildlife reserves, and ethical questions raised in relation to the recreational role of animals, for example as components of zoos and particularly circuses (e.g. Duffus and Dearden, 1990; Tudge, 1991; Shackley, 1996; Macnaghten and Urry, 1998).

Hall and Brown's (1996) conceptual exploration of a tourism research framework proposed that employment of the concept of welfare in relation to and of the elements involved in tourism can provide a means of illuminating tourism's nature and impacts. They argued that the welfare of one set of participants in the tourism experience will be influenced by efforts to increase that of others, and suggested ways in which the conceptual debate surrounding the nature and impacts of tourism development could be enhanced by adopting a welfare focus and organising framework. This can assist extending the frame of reference beyond that of human participants to encompass the welfare of, and ethical considerations concerning, those non-human animals most explicitly involved in providing tourism and recreation experiences for humans.

Employment of appropriate interpretational tools would appear to be a key element in the practical application of such perspective. Interpretation can transform experience by (humans) trying to represent interaction from the animal's perspective and by regarding the animal as a 'partner' for interaction, rather than as an object of curiosity and manipulation. This implies a stronger emphasis on three dimensions: (a) regarding animals as individuals, (b) the need for 'appropriate communication' between human visitor and animals, and therefore, (c) the need for less intrusive behaviour of the human visitor: for example, walking through wildlife areas rather than driving or using other mechanical means of motion. This requires visitors to be able to understand animals' actions and expressions and adjust responses accordingly – for example not staring directly head-on at a horse or at primates.

In the context of the interaction between humans and animals at visitor attractions, there appear to be two types of issue central to a welfare focus: (a) attitudes and responsibilities towards the role of animals in captive entertainment; and (b) the well-being of the ecological relationships enmeshing humans and animals and the environment within which their interaction takes place. Taking a somewhat utilitarian scientific approach to the definition of 'animal welfare', Broom (1992: 90-91) argued that the term referred to an animal's 'state', particularly its ability to cope with its environment. Such welfare can be measured scientifically on an individual basis and can be used to inform ethical decisions. Within this perspective, animal welfarists seek the best possible conditions for, and treatment of, animals. They tend to be opposed to circuses and factory farming. By contrast, the animal rights movement has had a more radical agenda (Regan, 1988), which has flowed primarily from the concept of universal justice. Both groups agree that the use of animals as objects of human recreation is outdated and often barbaric.

Arguments supporting the employment of animals for human recreation tend to emphasise that they are providing both entertainment and education, or are safeguarding traditional values and/or have conservation (and thus implicitly 'scientific') value (Clough and Kew, 1993: xiii, 3). Enhanced experience can be provided, particularly for children, from interacting with both tame and relatively wild creatures (e.g. Katcher and Beck, 1988), whereby a sense of respect, understanding and compassion can be generated. Conversely, if animals are seen to be 'exhibited' in poor conditions and/or manifest aspects of behaviour which reflect boredom, frustration or pain, the human experience can be negative and possibly traumatic. But of course, there is the potential threat, however small, of the transmission of disease, or other physical harm, from animals to humans (with the threat usually greater for children), through interaction at visitor attractions.

Animal-based Attractions

In his earlier work, Orams (1996b) proposed a simple three-fold classification – 'captive', 'semi-captive' and 'wild' of animal-based visitor attractions related to the extent to which animals are encaged and managed by humans. Duffus and Dearden (1990) and Reynolds and Braithwaite (2001) have viewed the relationship between animals and humans in terms of a distinction between consumptive and non-consumptive use (an explicitly human-centric evaluative stance). Non-consumptive wildlife-orientated recreation (NCWOR), is defined as 'a human recreational engagement with wildlife, wherein the focal organism is not purposefully removed or permanently affected by the engagement' (Duffus and Dearden, 1990: 214). Reynolds and Braithwaite (2001) suggested a three-fold categorisation of attraction contexts reflecting different levels of visitor access to animals: 'general access wilderness areas', 'limited access to rare/endangered animals', 'contrived experience of animals in zoos, circuses and aquaria, or with pets or other domestic species'.

With a view to place a greater emphasis on animals' interactive roles and the role of interpretation in enhancing mutual experience, Table 7.1 suggests that existing classifications could be extended.

Table 7.1　　Experiential classification of animal-based attractions

Category	Description	Experiential keywords
Wild animals for discovery and viewing in natural or semi-natural environments	a) Animals in 'natural' or semi-natural environments, at a distance from visitors, with little tactile contact. An underlying attraction may be the potential 'danger', 'wildness' and 'threat' of such animals, needing appropriate animal and visitor management schemes.	a) Distance viewing, photography, excitement, uncertainty, intrigue, thrill, unique, disappointment, danger, few people, privileged.
	b) Semi-natural environments – wildlife parks, wildlife rehabilitation centres, bird sanctuaries, managed country estates, large urban parks. As (a), but on a smaller scale and more managed. 'Wildness' is reduced, but the experience of visitors should remain of a similar character.	b) Viewing, photography, increased chance of closer viewing, anticipation, safety, timed feeding.
Wild animals as 'objects' for exhibition in a (captive) artificial environment	A more explicit form of animals as exhibition objects, often 'arranged' according to genus, ecological niche or geographical origin: zoos, aquaria, dolphinaria, butterfly centres and 'wildlife' centres. Little adventure or discovery is involved: visitors are separated from animals by physical barriers (fencing, glass walls, bars), which allow predictable and close viewing without risk. Animals are presented as entertainment and education. Conservation and rehabilitation as primary goals of many zoos and wildlife centres require animals to be kept in enclosures which better simulate their natural environment and help to educate visitors about their natural state.	Close viewing of many species in a small area; many people; not a unique experience; entertainment, education, information, conservation, some native but generally exotic, specialist species or large array of species.

Domestic/tame animals for interaction and discovery (captive and semi-captive)	Aimed at children, assuming such animals to be 'safe' (in their behaviour and from communicable disease), 'cuddly' and 'attractive'. They may be familiar to children through media images, but not normally through daily contact. On-farm 'presentations' as leisure activity, riding, children's farm, animal sanctuaries, animals providing transport at an attractions are included here.	Close viewing, touching, feeding, watching displays, husbandry, management, education, safety, friendly.
Wild animals for sport-based recreation	Animals used as targets for shooting, hunting, racing, or fishing. Some, such as foxhunting, may be justified on non-recreational grounds, such as the claim for pest control. Animals may be native or exotic, wild or semi-captive. Recreational game hunting usually requires prescribed legal and environmental circumstances, and as such, is often a collective activity. By contrast, fishing may be a solitary pursuit.	Diverse experiences: excitement, potential danger and sociability, to relaxation and solitude. Identity/social status symbols.

Visitor Perceptions and Expectations of Animals in Different Interactive Roles

Growing evidence suggests that the more 'control' the human visitor exerts over an animal, the more the visitor is inclined to humanise the animal and incorrectly interpret the animal's needs and feelings. Such an anthropocentric approach can lead to an unnatural (and probably uninspiring) experience for the visitor who may be left with little understanding either of the animal's behaviour in relation to the visitor gaze or of the animal's welfare needs.

Ironically, Reynolds and Braithwaite (2001: 37) employ the term 'control' when discussing methods for managing human-animal interactions. *Physical control* employs tangible separation, employing a guide or other forms of barrier external to the visitor. *Intellectual control* is seen as the level of expert knowledge transmitted by the guide or other interpretation mechanism.

But good interpretation is less about control but more about opening up – understanding, empathy, imagination: it is about stimulation rather than constraint. Through 'appropriate' interpretation for human-animal interaction, therefore, anthropocentric notions of 'control' can be superseded by objectives of 'interaction as partners' and 'communication', serving to convey both the 'otherness' of the animal and the uniqueness and meaning of interaction with it.

It is clear that the level of understanding generated can be strongly reflected in levels of visitor satisfaction – of clear importance for the viability of the attraction in terms of generating repeat visits and word of mouth recommendation. The impact of interpretation may be difficult to isolate from other variables such as educational level of visitors, communication with previous visitors and motivation levels (Reynolds and Braithwaite, 2001).

Value-added experience enhancement through an interpretation for visitors of what is 'appropriate interaction' is likely to differ substantially from the forms of information and 'education' – about the animal's habitat and statistics on its survival as a species – which are usually provided at such captive animal attractions as zoos, farm parks and wildlife parks.

Appropriate interpretation can emphasise the experiential quality of directness of personal contact and the need of the visitor to learn to interpret the animal's actions and expressions correctly. Two different (body) languages – of the animal *and* of the visitor – need an interpreter, and the interpretation of those languages has the potential to enhance substantially both the visitor experience and the dignity of the animal.

Experience Through Interpretation

As indicated in Chapter 5, interpretation increasingly is being adopted as a major and important tool for recreation management particularly in protected or sensitive areas, and may be employed in conjunction with codes of conduct (e.g. UNEP/IE, 1995; Mason and Mowforth, 1996). A growing recognition of the need for 'education' in tourism generally has resulted in the transmission of information in attempts to offset potential conflicts and negative impacts, often via the production of codes of conduct, many of which have had only limited effects (Mason and Mowforth, 1996). Means of persuasive communication are better developed in such contexts as health education and in advertising.

The potential for interpretive materials to make places and experiences accessible to tourists is well-documented (e.g Tilden, 1957; Moscardo, 1996; Orams, 1996b). Interpretation is the process by which information can be imbued with meaning – provoking thought, creating links, and communicating to people a sense of understanding and appreciation. In short, it is recognised as having the potential to enhance enjoyment (Ham, 1992), help develop an appreciation of sense of place (Stewart *et al.*, 1998), educate, challenge and provide insight (Tilden, 1957), facilitate attitudinal or behavioural change (Moscardo, 1996; Schänzel and McIntosh, 2000), relieve crowding and congestion (Moscardo, 1996), and assist in the management of nature-based tourism (Orams, 1996a, 1996b).

In tourism, the role of interpretation has tended to be harnessed less for the purposes of education than for entertainment. Interpretative materials and experiences now form the core of the tourism product, and are evident in a range of visitor attractions, for example in heritage and countryside centres. New leisure forms offer learning experiences staged as themes to provide entertainment by interpreting reproductions of the past, complexities of the present, and concepts of the future (Rojek, 1993: 146).

Behaviour does not change without a preceding change in attitude (Fishbein and Manfredo, 1992). In their turn, attitudes can be influenced by access to new information, appeals to emotion, and/or pressure from others. In theory, therefore, it should be possible to influence visitors' attitudes and behaviours in their interactions with animals by applying principles of effective interpretation to visitor management processes.

Most tourism-related research undertaken into the potential of interpretation relates to the natural environment. Of those studies that have been conducted, some have been in response to environmental concerns (Moscardo, 1996; Orams, 1996b, 1997; Stewart *et al.*, 1998) and others have arisen out of a recognition of the importance of interpretation techniques in responding to the needs of eco- and nature-based tourists. For example, Stewart *et al.* (1998), in projecting a context of *sense of place*, ask whether interpretation has the capacity to take visitors one step beyond wherever they happen to be when they arrive at a site so that empathy or 'care' is developed towards the conservation of that place and sequentially to other places throughout the world. The authors argue that well conceived and managed interpretation can bring about a significant attitudinal difference and may have a cumulative effect encouraging the desired result of empathy with surroundings. Schänzel and McIntosh (2000) for example argue that 'increased understanding' gained during a visit to the Penguin Place on the Otago Peninsula in New Zealand may lead to greater environmental awareness off-site. Can this development of a *sense of place* be translated for interaction with a particular animal as a *sense of species*? Does effective interpretation have the capacity to take the visitor one step beyond their existing conscious relationship with an animal so that empathy or 'care' is developed towards the behaviour and welfare of that animal and sequentially to that of other animals?

For successful value-added experience of human-animal interaction it is central that the animal's needs and motivations are an essential element of interpretation: interpretive materials must contain communication strategies that create links between visitor and, in the case of animals, species, and enable the visitor to establish personally meaningful connections within the interpretive experience while maintaining, and preferably enhancing, the dignity of the animal. This requires material to be targeted at different types of visitor, their social settings and the levels of attention that can be achieved (Mason and Mowforth, 1996; Ballantyne *et al.*, 1998).

Interpretative Material as Persuasive Communication

If, as Urry (1995: 188) believes, tourism is a conscious state in which conventional calculations of safety and risk are disrupted, tourists may be reluctant to conform to the requirements of behavioural codes that may echo the constraints of everyday regulated existence and which may therefore be incongruent with their leisure aims. Thus one of the most important benefits reported at the Penguin Place on the Otago Peninsula was the enjoyment gained from being able to view wildlife at close range, in contrast to everyday experience (Schänzel and McIntosh, 2000). Whilst visitor management techniques may be employed to reduce the impacts of such demands, an interactional partnership approach to interpretation will rigorously test the limits of interpretative materials designed to restrict visitors' experiences even if only at specific times. The influences of the leisure context add a further dimension to interpretation. The further removed from visitors' expectations the experience of interacting with animals may be, the more intensely the interpretative materials will be tested.

For leisure and tourism contexts, importance and enjoyment are essential pre-requisites to sustained involvement (Dimanche *et al.*, 1991). One purpose of effective interpretation should be to provide inter-active modes of delivery of relevant and provoking material which can attract and engage visitors and increase learning (Moscardo, 1996), thus contributing to a mindful visitor state and potentially, therefore, to an enhanced sense of species. The assumption is that the greater the extent of involvement in the experience, the more likely a visitor will be open to attitude and behaviour change when exposed to new information about it.

In relation to values and prevailing subjective norms, the theory of reasoned action (Chapter 5), suggests that many behaviours are determined by intention. It posits that the more a person believes that a certain type of behaviour will lead to positive outcomes (or prevent negative ones), the more likely that person will develop a belief to influence their behaviour. For example, if someone believes strongly that by not staring at a primate its welfare can be assisted, that person will not do so even though others may.

A less strongly held belief in the importance of not staring at a primate may render a behavioural decision more open to influence from 'normative' behaviour – the fact that staring at a primate is commonplace and that, for example, zoo arrangements may appear to encourage it and thus to indicate that it is not unacceptable. If volitional control is low and individuals feel that their contribution will make no difference, they may change their attitude but not necessarily their behaviour. Thus, for example, they may believe that staring at a primate is inappropriate but will do it anyway because everyone else seems to do so.

Success in effecting a change in a given intention (and possibly behaviour) is therefore more likely if one first alters the attitudes and/or subjective norms that correspond to it (Fishbein and Manfredo, 1992: 35). Thus, in order to reduce staring at primates, interpretative material might deliver two messages: one to influence personal attitude (the way an individual feels about being personally responsible for staring at a primate), and one to influence normative issues (the way an individual feels about primate welfare generally).

The number and complexity of processes involved in the development of good interpretive materials may explain why their application is still evolving. An understanding of the social and psychological processes involved is not widespread within the professions charged with visitor management. The continued absence of interpretive programmes in many tourism development plans may reflect the future-orientation and intangibility of outcomes that render them difficult to measure and near impossible to justify in economic terms. Yet evidence of the experience economy suggests that enhanced experience through an enrichment of human-animal interaction and of an understanding of the behavioural and welfare needs of the animal in order to inform such an interaction, will make for an entrepreneurially and ethically successful visitor attraction.

Conclusion

A rapidly increasing recreational demand for personal interaction with animals is intersecting with a growing awareness of our potential to disrupt the lives of animals and to negatively affect their welfare. The employment of animals as recreational 'edutainment' raises a number of potential conflict and contradictions which the demands of the experience economy may act to heighten. Yet appropriate interpretation, as part of a wider visitor management policy, can help to enhance and enrich human – animal interaction and experience for both. As value-added mediation, interpretation has the potential to emphasise the experiential quality of directness of personal contact, and in the context of animal attractions can assist visitors to learn to interpret animals' actions and expressions.

There is a clear need to better understand the nature of the visitor-animal relationship in the various types of captive animal-based visitor attractions, for ethical, experiential and business reasons. Within such a context, this chapter has promoted a number of arguments. First, that the experience of human interaction with an animal or animals in a range of recreation-related settings can be a uniquely rewarding experience. Second, that such an experience can be the main objective or a supplementary experience to other recreational activities within a visitor attraction. Third, that use of appropriate interpretative materials in framing the interaction may substantially enhance the experience through providing the visitor with an insight into the behaviour and welfare needs of the animal(s), permitting the visitor to behave in relation to the animal(s) in a more appropriate and aware manner, with the consequent likelihood of the animal's dignity and welfare being enhanced; and, last, that an entrepreneurially as well as ethically successful visitor attraction can result from the development of such attraction management tools.

Human interaction with animals has the potential to encapsulate all four realms of Pine and Gilmore's (1998: 102) conception of experience – entertainment, education, escapism and aestheticism. But only by incorporating into such experience an understanding of the behaviour and welfare needs of the animal, through the medium of appropriate interpretation, can such an interaction become a truly rewarding experience.

Implicitly this chapter has raised a number of questions which should be addressed by further research. In such contexts as evaluated in this chapter, who interprets? what is, and who judges, appropriateness? and how can the outcomes be measured: for animal welfare, for the visitor experience, and for an attraction's financial viability?

Acknowledgements

This chapter draws on research commissioned by the Scottish Executive Environment and Rural Affairs Department (SEERAD), and on research for the EU PHARE programme and the UK Government Foreign Office Know-How Fund programme undertaken in Central and Eastern Europe. Earlier versions were presented at conference gatherings in Rovaniemi, Finland and Jhansi, India, and the authors acknowledge the contribution of those participating in the discussions which followed the presentations.

References

Ballantyne, R., Packer, J. and Beckman, E. (1998), 'Targeted Interpretation: Exploring Relationships Among Visitors' Motivations, Activities, Attitudes, Information Needs and Preferences', *Journal of Tourism Studies*, vol. 9, pp. 14-25.

Broom, D.M. (1992), 'Welfare and Conservation', in R.D. Ryder (ed.), *Animal Welfare and the Environment*, Duckworth, London, pp. 90-101.

Clough, C. and Kew, B. (1993), *The Animal Welfare Handbook*, Fourth Estate, London.

Curry, B., Moore, W., Bauer, J., Cosgriff, K. and Lipscombe, N. (2001), 'Modelling Impacts of Wildlife Tourism on Animal Communities: A Case Study from Royal Chitwan National Park, Nepal', *Journal of Sustainable Tourism*, vol. 9, pp. 514-29.

Dimanche, F., Havitz, M.E. and Howard, D.R. (1991), 'Testing the Involvement Profile (IP) Scale in the Context of Selected Recreational and Touristic Activities', *Journal of Leisure Research*, vol. 23, pp. 51-66.

Duffus, D. and Dearden, P. (1990), 'Non-Consumptive Wildlife-Oriented Recreation: A Conceptual framework', *Biological Conservation*, vol. 53, pp. 213-31.

Fishbein, M. and Manfredo, M.L. (1992), 'A Theory of Behaviour Change', in M.L. Manfredo (ed.), *Influencing Human Behaviour: Theory and Applications in Recreation, Tourism and Natural Resources Management*, Sagamore Publishing, Sagamore, pp. 29-48.

Hall, D. and Brown, F. (1996), 'Towards a Welfare Focus for Tourism Research', *Progress in Tourism and Hospitality Research*, vol. 2, pp. 41-57.

Ham, S. H. (1992), *Environmental Interpretation*, North American Press, Boulder CO.

Katcher, A.H. and Beck, A.M. (1988), 'Health and Caring for Living Things', in A. Rowan (ed.), *Animals and People Sharing the World*, University Press of New England, Hanover, pp. 53-73.

MacCannell, D. (1999) *The Tourist: A New Theory of the Leisure Class*, University of California Press, Los Angeles, 2nd edn.

Macnaghten, P. and Urry, J. (1998), *Contested Natures*, Sage, London.

Mason, P. and Mowforth, M. (1996), 'Codes of Conduct in Tourism', in C. Cooper (ed.), *Progress in Tourism and Hospitality Research, Volume 2*, Belhaven Press, London, pp. 151-67.

Moscardo, G. (1996), 'Mindful Visitors, Heritage and Tourism', *Annals of Tourism Research*, vol. 23, pp. 376-97.

Orams, M.B. (1996a), 'A Conceptual Model of Tourist-Wildlife Interaction: The Case for Education as a Management Strategy', *Australian Geographer*, vol. 27, pp. 39-51.

Orams, M.B. (1996b), 'Using Interpretation to Manage Nature-Based Tourism', *Journal of Sustainable Tourism*, vol. 4, pp. 81-94.

Orams, M.B. (1997), 'The Effectiveness of Environmental Education: Can We Turn Tourists into Greenies?', in C. Cooper (ed.) *Progress in Tourism and Hospitality Research, Volume 3*, Belhaven Press, London, pp. 295-306.

Pine, B.J. and Gilmore, J.H. (1998), 'Welcome to the Experience Economy', *Harvard Business Review*, July-August, pp. 97-105.

Regan, T. (1988), *The Case for Animal Rights*, Routledge, London.

Reynolds, P.C. and Braithwaite, D. (2001), 'Towards a Conceptual Framework for Wildlife Tourism', *Tourism Management*, vol. 22, pp. 31-42.

Rojek, C. (1993), *Ways of Escape: Modern Transformations in Leisure and Travel*, Macmillan Press, London.

Schänzel, H.A. and McIntosh, A. (2000), 'An Insight into the Personal and Emotive Context of Wildlife Viewing at the Penguin Place, Otago Peninsula, New Zealand', *Journal of Sustainable Tourism*, vol. 8, pp. 36-52.

Shackley, M. (1996), *Wildlife Tourism*, International Thomson Business Press, London.

Stewart, E.J., Hayward, B.M. and Devlin, P.J. (1998), 'The "Place" of Interpretation: A New Approach to the Evaluation of Interpretation', *Tourism Management*, vol. 19, pp. 257-66.

Tilden, F. (1957), *Interpreting Our Heritage*, University of North Carolina Press, Chapel Hill NC, 3rd edn.

Tudge, C. (1991), *Last Animals at the Zoo*, Hutchinson Radius, London.

UNEP/IE (United Nations Environment Programme, Industry and Environment) (1995), *Environmental Codes of Conduct for Tourism*, UNEP, Paris.

Urry, J. (1995), *Consuming Places*, Routledge, London.

Wemelsfelder, F. (1997), 'Investigating the Animal's Point of View. An inquiry into a Subject-Based Method of Measurement in the Field of Animal Welfare', in M. Dol, S. Kasanmoentalib, S. Lijmbach, E. Rivas and R. van den Bos, (eds), *Animal Consciousness and Animal Ethics*, Van Gorcum, Assen, pp. 73-89.

Wemelsfelder, F. (1999), 'The Problem of Animal Consciousness and its Consequences for the Measurement of Animal Suffering', in F. Dollins (ed.), *Attitudes to Animals: Views in Animal Welfare*, Cambridge University Press, Cambridge, pp. 37-53.

Wemelsfelder, F. (2001), 'The Inside and Outside Aspects of Consciousness: Complementary Approaches to the Study of Animal Emotion', *Animal Welfare*, vol. 10, pp. 129-39.

Young, K. (1998), *Seal Watching in the UK and Republic of Ireland*, International Fund for Animal Welfare UK, London.

Chapter 8

Authenticity – Tourist Experiences in the Norwegian Periphery

Mette Ravn Midtgard

Introduction

Most Norwegians like to believe that Norway's scenery is the most extraordinary, beautiful and scenic in the world. This belief has its roots in the strong nationalisation process, which began around 1800 and culminated with national independence in 1905. Norway had been under Danish and Swedish rule for 400 years, and the ideological foundation and central symbols of the Norwegian separatist movement were nature, the countryside and the farmer. Authors, painters and composers also spread romanticism and a national idyll, leaving deep impressions on Norwegian national identity. In addition, the Nordic countries experienced late industrialisation and urbanisation, mostly during late 19th and 20th centuries (Nedrelid, 1991; Sandell, 1991). This partly explains why Norwegians have kept strong emotional bonds to pre-industrial and rural communities. The ancestral, spatial and cultural relationships to the countryside and the natural environment are said to be reflected in their leisure activities (Kaltenborn, 1993).

Romanticism has influenced perceptions of landscapes and nature, as well as cultural values, tastes and preferences relating to the natural environment. In addition to this cult of nature, romanticism brought about a nostalgic interest in the social and cultural aspects of rural communities, including history, heritage, simple and natural lifestyles. Often romanticists are inclined to dislike the rapid modernisation and technological transformation of rural landscapes and communities because they do not fit in with idealised images of nature and rural scenery. 'Untouched nature' and 'authentic' rural communities become 'green dream places' for victims of the far-reaching industrialisation and urbanisation of society. Areas that have escaped this violent process of modernisation and technological development become new objects of worship and places of pilgrimage (Wang, 2000). The simple, idyllic, and authentic quality of the 'green dream-place' makes it a contrast to an urban world of over-complexity and pollution. Thus, this antagonism generates a great interest in nature tourism. Romanticism is characterised as 'imagination' and 'feelings', and includes a passion for landscapes and scenery. Nature was considered a source of pleasure, and love of nature in turn paved the way for nature tourism. The 'place myth' that writers, painters and visitors contribute to has played an important role in the development of the taste for nature (Urry, 1995). Representations that characterise a region, a local destination or attraction are continuously being produced and reproduced by the visiting tourists and the tourism industry (Saarinen, 1998).

Tourism literature often emphasises the tourists' quest for authenticity, which means that modern tourists are in search of unspoilt nature, landscapes and local cultures with a distinct regional or local character. The tourists' nostalgic interest in traditional communities and in different and often less complex cultures can also be seen as a part of this major trend. The purpose of this article is to show how independent tourists in the Lofoten archipelago express their tourist experiences according to the term authenticity. Lofoten is a major tourist destination in Northern Norway, and one of the *brands* in Norwegian tourism marketing. Yearly, there are 250,000 registered overnight stays in the region, 70 per cent of which are holidaymakers. Additionally, 500,000 passengers pass through the archipelago on the coastal steamer every year.

Method

A qualitative approach was used in order to ascertain how the natural and authentic aspects of fishing villages and their surroundings were perceived and interpreted by independent tourists who spent a minimum of two nights on the islands. The interviews took place in the summer 1999.

The methodological approach follows Geertz (1993), whose point of departure is that knowledge about the object of study comes through respondents' presentation of their own practices, which in turn have been shaped by memory and personal interpretations. Knowledge cannot be based on the researcher's personal observations of concrete practice, because the researcher needs the help of the involved parties in order to understand the meaning of the signs and the practices he or she observes and interprets. The researcher's rendering and understanding depends, in a fundamental way, on the interpretations of the informants themselves. This implies that social and cultural meaning is constructed through active interaction between the researcher and the respondents. The result of this process of interaction is thus something constructed rather than something 'discovered'. At the same time, the fact must never be concealed that, by recreating a social reality, we are also simplifying it and risk distorting it (Geertz, 1983). Qualitative studies and the data collected during them are in reality interpretations of how the people studied present their own reality and practice and how aware they are of them (Geertz, 1993).

A total of 28 interviews were carried out, and in 18 of these two or more respondents in the same tour group took part. In all 48 respondents were interviewed. The interviews were taped and transcribed in full. The interviews took from 45 minutes up to 1½ hours. An interview guide was used in order to cover the topics that were seen as crucial in the project.

The Lofoten Islands

The Lofoten Islands are located off the coast of Northern Norway. The alpine mountains create a sharp contrast to the surrounding ocean. For many centuries, the people of these small communities and close-knit local villages lived off the rich fisheries in the area. Over the past decades, however, increasing numbers of tourists have travelled to enjoy the spectacular natural environment. Many of them stay in local fishermen's cabins ('rorbu') during the summer season. A lot of these old cabins have now been refurbished and transformed into tourist accommodation. In other areas, new and modern 'rorbu' tourist resorts have been built, in order to recreate the atmosphere of a typical fishing village of the area. The wooden cabins are located by the water and are built in the traditional style and in bright colours.

This chapter is based on a qualitative study carried out among individual tourists in three Lofoten locations: *Nusfjord, Henningsvær* and *Nyvågar*. Nusfjord has been preserved as the fishing village used to be a few generations ago, representing 'objective authenticity'. Visitors in Nusfjord perceive the village as old and genuine, with a distinct character. They find the place very charming and idyllic with a physical setting well protected from the sea. Henningsvær is a lively community with a mix of new and old buildings and both fishing and tourist activities. In Henningsvær tourists appreciate the scenery, the actual location of the community and the variety of services offered. Nyvågar is a result of modern tourism development. The respondents in Nyvågar appreciate being near the sea and high quality tourist facilities. Taken together, Nusfjord can be seen as the 'preserved' case, Henningsvær as the 'organically developed' case, and Nyvågar as the case 'generated by the tourism industry'. The natural surroundings are fairly similar in all three locations.

'Authenticity' and Tourism Development – Theoretical Approach

There has been much debate about the meaning and importance of authenticity in modern tourism (see for instance Redfoot, 1984; Pearce and Moscardo, 1986; MacCannell, 1989; Cohen, 1995; Wang, 1999; Dann and Jakobsen, 2002). According to Ning Wang's analytical approach, there are three main types of authenticity: objective, constructive and existential authenticity.

a) *Objective authenticity* is based on absolute and objective criteria (the museum is an example). Bruner (1996) defines four different meanings of authenticity, as objectively understood. First is an authentic reproduction as like the original as possible. Secondly it could be an object that is historically accurate and conformed or thirdly, it might be original, as opposed to a copy. The fourth category is defined as a certificated or branded good, by someone authorised. According to this classification Wang claims that this implies a simplification of the original (1999: 354). He writes '[...] in this sense, if mass tourists emphatically experience the toured object as authentic, then, their viewpoints are real in their own right, no matter whether experts may propose an opposite view from an objective perspective' (Wang, 1999: 355) with reference to Cohen (1988).

b) *Constructive authenticity* refers to the authenticity projected onto objects by tourists or tourism producers in terms of imagery, expectations, preferences, beliefs etc (Wang, 2000: 49). This may be described as symbolic authenticity or authenticity as a result of social construction. Whether or not tourists themselves perceive an object as authentic is the crucial question here, and it is also possible to assume that both *objective* and *existential authenticity* are socially constructed. Both experiences and objects can be authentic and they constitute each other, while it is perhaps necessary to regard the authenticity of toured objects as symbolic authenticity. Either way, this implies the social construction of both forms of experience (actual experiences versus objects). It is therefore difficult to make absolute distinctions between the different forms of authenticity. The experience of enrapture described as existential authenticity might even be a socially constructed happiness, and so might the idea of an authentic product; it might represent a restored myth. The effort to create fishermen as examples of objective authenticity are not included in the concept of constructed authenticity. The 'constructive transformations' take place inside the tourist mind.

c) *Existential authenticity* is defined by a so-called existential state of being. The feeling is associated with an Eros orientation, which means that emotions, feelings and spontaneity are in the foreground (Wang, 2000) rather than reason or self-constraint (Logos). In such subliminal experiences, the tourists feel themselves to be much more authentic and expressive than in their everyday lives. The authentic experience implies complete involvement in a situation or social interaction. Simmel called this experience *Erleben,* a state in which the 'I' (the individual) and the perceived surroundings melt together (Cederholm, 1999). It is an holistic feeling of being completely at one with the natural environment or the social and cultural context. This state of mind cannot be sustained for a very long period, and after an intense and affective experience, the person involved will create a consciousness of the experience and start to reflect on the extraordinary situation. The experience then becomes objectified, something that can be expressed and shared with other people.

Existential authenticity is similar to the experience of being enraptured. Rapture presupposes a complete emotional involvement in a tourist destination with a unique character. Such a strong involvement in a location may be the result of a particular interest in the scenery, of fascination with nature, the history of the destination, its social and cultural assets, and the myths associated with the place. A certain amount of time in the place is a prerequisite for such a search for an in-depth experience. In this state, it is possible for the visitor to grow into the natural or local social rhythms. Existential authenticity can be attained when the individual is far away from everyday life and open to new ideas and impressions and there is a true emotional fascination with the place being visited.

In the modern world Logos dominates in most institutional settings, and Eros is more or less constrained. One might see the quest for the authentic experience as a reaction to the mainstream order of modern society and the need to create a space outside the dominant rational order of everyday life. Tourist experiences can offer such a space beyond the borders of daily norms and routines. Logos is maybe best associated with objective authenticity (Wang, 2000).

Travel Motivation and Images of Lofoten

Generally, visitors to Lofoten, or at least our respondents, had a fairly similar understanding of the concept of authenticity; authentic to them was *something* that still looks as it did in another epoch. In other words, the authentic had to include an element of tradition, and when respondents referred to objects they found authentic, these were largely the same ones that were advertised in brochures and photographs of Lofoten: fishermen's cabins, fishing smocks, nets, quays and so on. Many other aspects also contributed to the tourists' experiences of authenticity. The relatively simple way of living for the independent traveller was highly appreciated and enjoyed by the respondents, and this type of vacation was seen as liberation from the restrictions and conventions of everyday life. The effect of staying in small-scale, transparent local communities or tourist resorts put people in a different mood and opened up their senses. A certain amount of time is necessary to accept and adopt the slow pace and local ways of doing things.

When asked why they visit Lofoten, the respondents usually answered that they wanted to experience the scenic beauty of the islands (Prebensen, 1999; Haukeland, 1999; Dann and Jakobsen, 2002; Jacobsen *et al.*, 2002). This reason was far more important than any other, and was often the only reason visitors gave for visiting the archipelago. Other reasons were secondary or additional. Typical secondary motivations were outdoor activities, peace and tranquillity, and the chance to enjoy a simple lifestyle and experience the slow pace of local life. Norwegian respondents often emphasised socialising with the locals, while foreigners mentioned local culture and atmosphere. Many said that vacationing in a fisherman's cabin gave them a chance to rest and relax with their family. The widely-held view that the area is not over-crowded with other tourists was also mentioned in this context.

Respondents were asked about their knowledge and perceptions of Lofoten before they arrived in the area. At the outset the respondents' overall image varied a lot, but most associations had to do with natural features like the steep mountains surrounded by the wide open ocean and fjords, the unique light and changing weather conditions. They envisioned a region with great visual contrasts. Lofoten also seemed to be well known among our respondents for its old fishing history and an image of picturesque fishing villages with fishermen's huts and small fishing harbours. These observations, however, were probably reinforced by actual visitor experiences as the interviews took place some days after their arrival.

The Authentic Lofoten – Different Concepts and Experiences

This section uses Wang's (2000) theoretical approaches and concepts to classify the visitors' statements concerning their tourist experiences in Lofoten, with the proviso that they have been interpreted by the researchers, particularly emphasising the value of nature in tourism in peripheral areas.

Objective Authenticity

The visitors said that Lofoten had an aura of authenticity that came from the original character of the area and that this character had not been greatly transformed by the recent tourism development. Both natural and man-made elements contributed to this impression of objective authenticity. The interviews revealed that the dramatic scenery is what gives Lofoten its unique character; and the steep mountains were emphasised in particular. The respondents saw the contrasts of the scenery as a distinguishing feature:

> The contrasts of ocean, northern lights, nature, and mountains, and in particular the great contrast between the mountains and the ocean...

Nature was seen as something that had not yet been marked by human activity, as a German respondent pointed out, when asked about her/his associations with the authentic Lofoten:

> I believe it is mostly connected to clean and unspoilt nature. In Germany, for instance, nature is much dirtier.

Many respondents emphasised the particular value of the original character of the fishing villages. For instance, another German tourist in Henningsvær stated:

> I think it is very old, and it is a good thing. What I experience here is a true atmosphere. You sit here and you know that the fisherman really used to live just here. There is something in these walls that you cannot recreate when it is a new thing, and that atmosphere is what makes this place.

And the same type of reflection is revealed by visitors in Nusfjord:

> People know that Nusfjord is a preserved place. And it gives a special feeling to experience some of the past, to see how it was many years ago. Our Swedish friends are totally satisfied with the place. It is as primitive as possible with a bucket for keeping the water inside and a toilet right on the quay. They find it perfect. The standard in our cabin is a bit too high. The one beside us is more characterised by a primitive interior – the bucket at the bench and one hot plate in the corner. We wouldn't mind that standard either.

Most of the respondents staying in original fishermen's cabins saw this type of accommodation as real and truly authentic, and as customers they seemed quite satisfied with the simple tangible good of these relatively primitive cabins. In general, the respondents also noticed a certain lack of local aesthetic values in different parts of Lofoten, and these impressions were seen as stemming from the historically simple and meagre social conditions in the area. These aspects may well reflect the true character of the local culture:

> The pink school building, or whatever it is – it is special. That reminds us a bit of indigenous people: Greenlanders, Laps, Eskimos... It is some kind of simple style formed by the poor living conditions. One slapdash of paint with those strong colours reflects some sort of missing aesthetic sense.

Constructive Authenticity

Constructive authenticity is the social construction of what is genuine Lofoten. This means for instance that a common expression is that the region is beautiful. This impression and expression is widely accepted now, probably socially constructed in the later decades. There are several writers and visitors to the region who expressed the opposite view in the 19th century and even in the 20th century. One famous writer who found Lofoten ugly is the philosopher Jean Paul Sartre (Beauvoir, 1992).

This expression captures what is meant by the term social constructed authenticity of Lofoten:

> It is impossible to describe the beauty; you just have to experience it!

Several respondents gave answers that indicated that the toured location, the preserved Nusfjord, for instance, had physical qualities that better gave them other experiences of authenticity. They can, for example, create their own histories about the place and the lives of the former inhabitants.

> When we are staying in a place for some time, then it is a trial period for experiencing the 'evocative mood' (Wahrnehmung); how the people really are, and how and under what atmosphere they have lived earlier.

Some of the respondents were concerned with the problem of making new products by reproducing copies from previous periods. Those who stayed in the older, original cabins supposed that it would be impossible to construct copies with the same authentic and unique character. Moreover, some of these respondents thought that new establishments could never be as harmonious as the older ones:

> It is impossible to imitate; you will miss some of the old – what smells old. This is a little bit of nostalgia.

In spite of the fact that some things are not necessarily original, the tourists may enjoy elements that are perceived as having a distinct local character. In Nyvågar, the relatively new and artificial 'village', the respondents pointed out that the architecture was typical of the region and harmonised with the natural surroundings. In Henningsvær and Nusfjord, the older buildings, quay structures and boats were seen as typical of Lofoten fishing villages. The visitors of each place, because of the location and the built environment, perceived the character of Nusfjord, Nyvågar and Henningsvær as typical of Lofoten. However, the respondents in Nusfjord and Henningsvær had a tendency to disassociate themselves from the type of accommodation found in Nyvågar. By contrast, some interviewees were critical of the objective authenticity of the villages as they were more or less transformed and arranged for visiting tourists, almost like a museum:

> Unfortunately it is no longer a living fishing village. It is just the memories in the houses.

This quotation refers to the toured object, while at the same time recognises some kind of internal subjective feeling.

Existential Authenticity

The experience of existential authenticity is identified when there is a strong degree of involvement and the respondent is excited about the true and authentic character of the place visited. The study supports the assumption that there is an element of existential experience among the visitors in Lofoten. A state of existential experience is reached when the surrounding world is perceived with all the senses and the visitor 'melts together' with the place visited (Wollan, 1999).

A couple of quotations might illustrate these absorbing experiences. A businessman from Southern Norway, for instance, reports:

> Just sitting by the window like I'm doing now and watching the people, the water, watching the movements, looking at the surroundings, is in itself an activity and experience that soothes the soul. The feeling you get here that you can turn off and gear down to idling speed is just so wonderful.

A family man expresses it this way:

> Just sitting here knowing that you are at peace with yourself, that there is nothing churning inside you. Just being able to sit there and enjoy yourself and let the hours go by, that's a fantastic experience.

A woman from Berlin shares her reflections on her and her travel companion's experiences in this way:

> We both take a chair and sit down in front of our house, we can sit in one place for three to four hours. And every five minutes we will see something different. The very way the light varies. The colours, yes, the colours are impressive.... The very cloud formations,

then again the blue, then dark blue, the shadows cast on the water, the reflection of the house across the bay. For instance, on Wednesday we sat for four to five hours on a single rock and never got bored! We also hardly spoke to each other. The animals and birds, too, the way they speak to one another, or the way we think they speak to one another. They call out to one another, and then we have to be quiet. The truth is, you have to listen to the seagulls etc. I can never have this at home. It is not at all possible.

Sensual experiences include breathing sea air and the smell of the seaweed and fish. The freshness of the air is also noticed. The sounds of running water, waves and birds singing belong to the natural world. Tranquillity is found in such a natural environment and is highly appreciated by the visitors as a break from an everyday life that is often reported to be demanding and stressful. Watches are often left at home, and the individual relates instead to the rhythm of the tide and the light conditions. The respondents often find peace and inner harmony by watching nature unfold.

My wish is to make a small fishing trip on the bare rock-face, to find peace and comfort, to just be sitting there and daydreaming.

There are examples of intimate testimonies of the genuine quality of peace and quiet, ambience and atmosphere, and the authentic character of the locality. Deeply felt experiences of both natural elements and the built environment may thrill the tourist in an absorbing way. These are qualities that are perceived even if some tourists claim that these values might be threatened as a result of sudden modernisation or rapid expansion of the tourist industry in the area. As shown in the quotations above, the importance of nature and the natural variations, together with silence and tranquillity is what visitors most often single out. They also mentioned what kind of conditions are needed for benefiting from these special emotions. Perhaps most important of all is the time spent at a location. People seeking this peaceful feeling obviously know they need time to settle down and be in harmony with themselves.

Authenticity and the Tourist Experiences in Lofoten – Central Findings

Objective authenticity is perceived when an environment is seen as genuine or original. The Lofoten respondents usually see the physical environment, such as the buildings, as authentic. They emphasise the landscape with all its contrasts, the relatively clean natural environment, the traditional fishermen's cabins, fishing boats, harbours and fish drying racks as elements that are perceived as unique and therefore difficult or impossible to copy. Tourists come with expectations that they actively seek to fulfil. Other studies show that it is less important whether these symbols have any relevance for today's society or reality in the specific historical background, as long as the image held by the visitors is experienced (Lyngnes and Viken, 1998). Lofoten also shows many of the marks of modern Norway and modernisation processes. When potential new industrial developments have been discussed locally, people have wondered whether new manufacturing facilities, for instance, would spoil a fishing village for the visitors. The impression I gained from the respondents was that it would not. Tourists go 'shopping' for sights, and enough of the old Lofoton still remains for them to find what they are looking for. If they stay in a modern hotel, for instance, they are likely to go and see an old fisherman's cabin so that they can 'tick off' that particular image. And if they notice a newer building or something else that does not fit their image of Lofoten, they tend to look the other way. The incongruent element is hardly recognised as part of their tourist experience. When asked they usually say that of course they expect to see signs of the modern world. They understand that the locals have to work and need modern conveniences. People playing the social role of tourists are nevertheless reflexive people.

The line between objective and constructive authenticity might be hard to draw. The Nyvågar resort is an example of a reproduction that is not experienced as authentic in the objective sense, but visitors for the above-mentioned reasons nevertheless appreciate it. Despite the absence of a local population in this holiday village, the presence of other tourists, the scenery and other symbols of Lofoten make the place attractive. These qualities represent constructed symbolic meanings but they cannot be categorised as constructive authenticity. The answers of the respondents cannot tell us conclusively whether it is objective or constructive authenticity they experience. The construction of symbols, as in objective authenticity, or the innate qualities of man-made and natural objects (nature fascination), as in existential authenticity, may in many cases be hard to differentiate.

The most striking finding was clearly the ability of the scenery to bring about existential authenticity. The experience of existential authenticity is characterised by a very strong degree of involvement and excitement about the authentic character of the place visited. Deeply felt experiences of both natural elements and the built environment may thrill the tourist in an absorbing way. When the surrounding world is perceived with all the senses, the visitor has a feeling of being one with the place visited, and an existential experience is reached.

This idealisation of the natural world and deeply rooted rural life is a cultural value that should be interpreted in light of the fact that these aspects are under pressure in contemporary society. Underlining the importance of these qualities and expressing anti-tourist attitudes can be regarded as an escape from and reaction to the mainstream order of modern society. The paradox is that the kind of attitudes to tourism demonstrated are both a result of modern life and a reaction towards modernisation of the natural scenery and rural periphery.

The presence of other tourists and a certain amount of tourist infrastructure are generally regarded as positive, even if some respondents prefer to stay apart from other tourists. A strikingly common feature is the restrictive attitude to further development of tourism in Lofoten. There is a strong wish that the villages or the village-like resorts should remain small scale, and those who seek authentic experiences are worried that the current touristification might be too dominating.

The interviews also revealed scepticism towards certain forms of organised tourism that might be classified as anti-tourist attitudes. They were clearly negative towards converting old fishing villages into pure tourist resorts without any other locally-based activities. This chapter opened with a presentation about Norwegian romanticism and nationalism. Most of the informants were Norwegians, and the material shows that there is still a strong element of romanticism among them.

Besides, the study reveals a deep interest among tourists in experiencing nature and scenery in a profound and multifaceted way. Closeness to nature is a striking feature of the old fishing villages and the new tourist resorts in the region. The built environment and traditional way of living in this peripheral coastal area also contributes to the experience. There is a varying degree of emotional involvement in experiencing the assets of the Lofoten islands. In the most deeply felt version it reaches a level of total absorption or what can, in Ning Wang's words, be described as an experience of existential authenticity.

Conclusion

The interviews support the assumption that there is a quest for authentic experiences among independent tourists in Lofoten. The tourists give priority to the natural beauty of the area, but also emphasise man-made or social and cultural aspects. A variety of qualities are associated with the region and the villages. Peace and tranquillity are found in the silence of the landscape, the peace and harmony that is felt by sensing nature's own sounds, movements and rhythms. The slower pace of life, informality, hospitality and frankness of the local people also help visitors relax. There is a feeling of safety and friendliness. By enjoying these environmental qualities, it is possible to find rest and relaxation in the three places that were examined. Ambience and atmosphere are related to the perception of the destinations as small with clusters of waterfront Lofoten houses against a magnificent natural backdrop. This feeling of ambience does not, however, require a strong element of involvement among the tourists, and respondents in all three case areas expressed this kind of experience.

The data revealed a search for 'green dream places' and nostalgic wishes to experience human environments that represent simpler lifestyles rooted in the rural past. It is reasonable to interpret these attitudes as an indirect support of alternative forms of tourism that are small-scale, ecologically sound, sustainable and sensitive, as they do not harm the natural environment or the authentic character of the local villages and communities. The tastes and values of individual tourists regarding nature and traditional rural communities parallel values often associated with romanticism. This movement has been the driving force behind essential parts of modern tourism, and these interests are maintained by the development of modern society.

Acknowledgement

The author would like to thank Jan Vidar Haukeland, Norwegian Transport Economic Institute, who assisted with the fieldwork and the writing of the initial conference paper.

References

Beauvoir, S. (1992), *Witness to my Life: The Letters of Jean-Paul Sartre to Simone de Beauvoir, 1926-1939*, Hamish Hamilton, London.

Bruner, E.M. (1996), 'Abraham Lincoln as Authentic Reproduction: A Critique of Postmodernism', *American Anthropologist*, vol. 96, pp. 397-414.

Cederholm, E.A. (1999), *Det Extraordinäras Lockelse. Luffarturistens Bilder och Upplevelser*, Arkiv förlag, Lund.

Cohen, E. (1988), 'Authenticity and Commoditisation in Tourism', *Annals of Tourism Research*, vol. 15, pp. 371-88.

Cohen, E. (1995), 'Contemporary Tourism, Trends and Challenges', in R. Butler and D. Pearce (eds), *People, Places, Processes*, Routledge, London.

Dann, G.M.S. (1976), 'The Holiday was Simply Fantastic', *Revue de Tourisme*, vol. 3, pp. 19-23.

Dann, G.M.S. and Jakobsen, J.Kr. (2002), *The Tourist as a Metaphor of the Social World*, CABI, Wallingford.

Geertz, C. (1983), *Local Knowledge*, Basic Books, New York.

Geertz, C. (1993), *The Interpretation of Cultures*, Fontana, London.

Haukeland, J.V. (1999), *The Fishing Village as a Tourist Attraction*, Paper for the 34th TRC–meeting in Vienna, 19th – 22nd March.

Jacobsen, J.Kr., Berit, G. and Haukeland, J.V. (2002), På veg mot Drømmeferien? Aktiviteter, Interesser og Opplevelser Blant Utenlandske Bilturister i Utvalgte Områder i Norge. *TØI report 575/2002*, Transportøkonomisk institutt, Oslo.

Kaltenborn, B.P. (1993), *Vårt Friluftsliv*, Temahefte 3, Norsk institutt for naturforskning, Lillehammer.

Lyngnes, S. and Viken, A. (1998), 'Samisk Kultur og Turisme på Nordkalotten', *Forskningsrapport nr. 8/1998*, Handelshøyskolen BI, Sandvika.

MacCannell, D. (1989), *The Tourist. A New Theory of the Leisure Class*, Shocken Books Inc., New York.

Nedrelid, T. (1991), 'Use of Nature as a Norwegian Characteristic. Myths and Reality', *Ethnologica Scandinavia*, vol. 21, pp. 25-36.

Pearce, P.L. and Moscardo, G. (1986), 'The Concept of Authenticity in Tourist Experiences', *Australian and New Zealand Journal of Sociology*, vol. 22, pp. 121-32.

Prebensen, N.K. (1999), *Trender og Utviklingstrekk i det Tyske Reiselivsmarkedet: et Samarbeidsprosjekt Mellom: Nortra, Color Line, Top Destinasjon*. Høgskolen i Finnmark og SND. HiF-rapport 1999:4, Alta.

Redfoot, D.L. (1984), 'Touristic Authenticity, Touristic Angst, and Modern Reality', *Qualitative Sociology*, vol. 7, pp. 291-309.

Saarinen, J. (1998), 'The Social Construction of Tourist Destinations', in G. Ringer (ed.), *Destinations: Cultural Landscapes of Tourism*, Routledge, London.

Sandell, K. (1991), *Re-creation or Creation – Outdoor Life; Re-creation of an Unsustainable Society or a Source of Inspiration for a More Sustainable Development?* Paper for 12:e nordiska symposiet fôr kritisk samhällsgeografi, Rosenön.

Urry, J. (1995), *Consuming Places*, Routledge, London.
Wang, N. (1999), 'Rethinking Authenticity in Tourism Experience', *Annals of Tourism Research*, vol. 26, pp. 349-70.
Wang, N. (2000), *Tourism and Modernity. A Sociological Analysis*, Pergamon, Amsterdam.
Wollan, G. (1999), 'Kulturarvturisme og Autentisitetens Etikk', in J.Kr. Jacobsen and A. Viken (eds), *Turisme – Stedet i en Bevegelig Verden,* Universitetsforlaget, Oslo.

Chapter 9

Rural Tourism and Film – Issues for Strategic Regional Development

W. Glen Croy and Reid D. Walker

Introduction

Many economies are investigating or encouraging the growth of new industries in areas that have been disenfranchised by technological advances and global production systems. De-industrialisation, globalisation and urbanisation are transformations that have occurred in many economies and are phrases that are commonly used, especially in the field of tourism development and rural tourism. These events in themselves have had dramatic effects on rural areas and tourism has regularly been chosen as a development mechanism to combat such events (Kotler, Haider, and Rein, 1993; Butler, Hall, and Jenkins, 1998; Hall and Kearsley, 2001; Roberts and Hall, 2001).

Similarly, events less associated with these global changes have dramatic effects on rural areas. The terrorist attacks on 11 September 2001 had dramatic effects on international tourism, as did the 2001 foot and mouth disease predicament in the United Kingdom. The foot and mouth disease especially identified a need to diversify the rural economic base. The destruction of agricultural industry had and will have major and long-lasting effects. The effects overflowed into the UK tourism industry, crippling another important prospect for rural areas. In this situation, as in other rural areas, residents must be offered a means to sustain a livelihood, with a view to once again re-establishing dominant industries.

The first step in re-establishing these dominant industries of agriculture and tourism is to introduce a positively appraised image. Implicit within the consideration of destination image management is the source of image. To manage image includes much more than the information that a destination can officially and directly supply. The general media plays a very large role in the creation and dissemination of destination image (Moutinho, 1987; Butler, 1990; Altheide, 1997; Fodness and Murray, 1999), particularly so in the initial stages of destination image formation (Gartner, 1993).

Further, tourism marketing, like all other forms of marketing, is affected by the changes the marketing industry undergoes. Such recent changes include the decreasing response of consumers to promotion, and the effects of technological advancement (Belch and Belch, 1998). Additionally, despite significant funding being allocated to tourism marketing, the desired results are often not forthcoming (Nielsen, 2001). Within this new environment, fictional media, especially that of the feature film, have grown. Marketers have sought new opportunities for promotion and have achieved this by 'redefining the notion of what an ad is and where it runs. Stealth messages are being woven into culture and embedded into movies and TV shows' (Belch and Belch, 1998: viii). The use of fictional media is advantageous as it provides market coverage and where 'messages are memorable enough to capture awareness and sustain the interest of people who do not have the immediate ability to travel' (Riley and Van Doren, 1992: 267).

This chapter discusses the use of fictional media in rural areas to develop a tourism identity and tourism industry. The fictional media of literary works are first exemplified, followed by a brief discussion of the use of television series, before discussing the focus of this chapter, the fictional media of feature films. The feature film industry is introduced, followed by a discussion on the impacts of post-production exposure and more particularly film-induced tourism. This is followed by the introduction of film in New Zealand and research that has identified predominately rural local authority and regional tourism organisations' appreciation of the role of film in diversifying and developing their economic base.

Fictional Media Tourism

The turn towards tourism in rural areas is often developed based on iconic events of the area or of historical links to the past and in contrast to lost authenticity of the urban world. In the attempt to attract tourists many rural areas are turning to their historic links in the literary arts to reinvigorate their image. There are many examples of this occurring in the United Kingdom and North America, as elsewhere around the world. The study of literary tourism has also increased (Table 9.1).

Similarly, the New Zealand town of Oamaru has developed a heritage trail after Janet Frame's life and writing in the area. The Orkney Islands, Scotland, have also developed a literary tourism site at the Round Kirk in Orphir, the site of the infamous unfair drinking chapter of the *Orkneyinga Saga*. These predominately rural areas have further developed their associations with authors and literary works as a tool of promotion, developing imagery and adding meaning to their areas in the production of a tourism industry, or at least a tourist attraction.

Nonetheless, many rural areas without these historical links to literary icons are investigating other methods to develop a similar association with the fictional media. Most notable is the development of book towns (Seaton, 1999), and book festivals (Small and Edwards, 2002). More recently this has been taken further with the promotion and attraction of television production to create a new association with possible visitors. In many cases this also creates a retrospective association through the focusing on a specific period from the past, highlighting the dichotomised attraction of rural areas in an increasingly urban world. These television series have in turn been used to promote the rural areas in which they are set. Examples of the effects of television series on tourism and the rural area have been noted also in the literature: *Northern Exposure* and the small town of Cicely, USA (Hanna, 1996), *Sea Change* and Barwon Heads, Australia (Beeton, 2000, 2001a, 2001b), and *Ballykissangel* and Avoca, Ireland (O'Connor, 2001).

Table 9.1 Studies of literary tourism

Author	Year	Location	Focus
Curtis	1981	Monterey Peninsula and Salinas Valley, USA	Steinbeck
Curtis	1985	Hannibal, USA	*Huckleberry Finn* and *Tom Sawyer*
Squire	1994	Near Sawrey, England	Beatrix Potter
Herbert	1996	Northern France	Marcel Proust
Squire	1996	Prince Edward Island, Canada	*Anne of Green Gables*
Muresan and Smith	1998	Transylvania, Romania	*Dracula*
Cater	2001	Zhongdian, China	*Lost Horizon*
Fawcett and Cormack	2001	Prince Edward Island, Canada	*Anne of Green Gables*
Herbert	2001	Laugharne, Wales Chawton, England	Dylan Thomas and Jane Austen
Muller	2001	Sunne, Sweden Vimmerby, Sweden	*Gosta Berlings* Saga, Selma Lagerlof, and *Pippi Longstocking*, Astrid Lindgren
Weir	2003	England, the Middle East, Australia, Myanmar, and North America	Neville Shute

Whilst the rural-ness of each of these areas is presented in the books and television series, it is the association with people and stories presented that attract people to the rural location, it is the setting in which to re-interpret the events and to become part of the lives of those depicted in print and on screen. The inducing effects of these authors, books and television series is in some cases immense, and whole regions have marketed themselves on the basis of these features, as in the noted literature above.

Nonetheless, more importance is now placed on the feature film in the world of fictional media and it now yields great power as a form of mass media. This greater power has also brought increased attention in the study of mass media as a depiction device of and in society, mainly within the bounds of cultural geography (Anderson and Gale, 1992). The impetus has increased as media have been identified as major vehicles of awareness and leadership (Coates, 1991). Consequently, critiques of the media have increased because of the representations of society and specific groups within society (Jackson, 1992). Media reinforce, provide and maintain the norms of society, and as such are a significant influence in the creation of images and perceptions of people, place, race, country and culture (Wilson 1996). Contained within the mass media is the growing importance of cinematic representation (Aitken and Zonn, 1994a). Aitken and Zonn (1994b: 5) exemplify this in 'the way spaces are used and places portrayed in film reflects prevailing cultural norms, ethical mores, societal structures, and ideologies. Concomitantly, the impact of a film on an audience can mould social, cultural, and environmental experiences'.

The Feature Film Industry

The significance of the feature film, not just as a form of media and culture, but also as an economic industry in its own right, is immense. To put this industry and specific films in perspective, the *Star Wars* series has a combined worldwide box office taking of almost US\$3.5bn *(The Numbers*, 2003), almost three times the regular budget for the United Nations (United Nations, 2000). As is implied with the large box office takings of feature films, the viewing of feature films is growing as a leisure activity (Tooke and Baker, 1996). For example, in New Zealand, with a population of just over 3.5mn, 6.1mn cinema admissions were sold in 1991. This grew by an average of 38 per cent per annum to 16.3mn admissions in 1998 (A C Nielson, 1999).

Internationally, movie going as a leisure activity has also increased substantially. Throughout the European Union cinema admissions increased substantially in the period 1990 to 1998, achieving a total increase of 38 per cent (Deiss, 2001). This trend is reflected in other countries around the world and exemplified in Cushman, Veal and Zuzanek's (1996: 239) study of leisure activities of various countries where they noted 'no one activity is common to all surveys reported, although television watching and going to the movies comes close'. This is further illustrated in Table 9.2, which shows cinema admissions for USA, the United Kingdom and Australia.

Table 9.2 Cinema admissions in the USA, United Kingdom and Australia (in millions)

Year	1990	1991	1992	1993	1994	1995	1996	1997	1998	1999	2000	2001
USA	1,189	1,141	1,173	1,244	1,292	1,263	1,339	1,388	1,481	1,465	1,421	1,487
UK	97	100	104	114	124	115	124	139	136	140	143	156
Aus	43	47	47	56	68	70	74	76	80	88	82	93

Sources: Motion Picture Association (2002), Australian Film Commission (2003), British Film Institute (2003)

Coinciding with the increasing access to viewing facilities has also been the access to the production facilities of films. Promotion of the creation of films has also grown, both for economic gain and as a growing form of art appreciation. Many governments are now actively encouraging the production of films as a record and representation of culture. For example the New Zealand Film Commission's mandate is 'to contribute to the creation of cultural capital in Aotearoa/New Zealand through popular feature films' (New Zealand Film Commission, 2000). These themes are reflected in national film commission mandates around the world, although at a local level there is a predominate focus on the economic impacts of the film industry, and nowhere more so than in Hollywood.

Hollywood, Los Angeles, has long been identified as the world centre of the film industry, and benefits California by more than US$20bn a year (*The Economist*, 1998). Hollywood has recently been joined by Mumbai (Bombay) in India, colloquially known as 'Bollywood' as a world centre of film production (Film New Zealand, 2001). Nonetheless, film production has undergone substantial geographic dispersal, to the point that sections of the USA film industry began to lobby their government to retain production in the USA. Runaway production, as this phenomenon is known, is costing the USA economy billions of dollars in lost earnings annually (International Trade Administration, 2001; Monitor Company, 2001), and the loss of an estimated 20,000 jobs in the American film industry (Weller, 1999). At the same time many film producers are now looking globally for production sets, for unique and picturesque locations, and because of internationally competitive production costs (International Trade Administration, 2001). The most common locations for runaway productions are in continental Europe, the United Kingdom, Canada and Australia.

Runaway productions are creating significant economic benefits for host regions. For the period 1996-2000, Sydney, Australia, gained A$80-100mn from Hollywood funded film production (Fitzgerald, 2000), and British Columbia, Canada, US$1bn (Hunter, 1999; Townson, 2000). Other areas of the USA are also reaping the benefits of film production, including Georgia attracting US$100mn (Thompson, 1992), Virginia US$72mn (Virginia Tourist Corporation, 2000), and Illinois, US$100mn (*The Economist*, 1998). The public sector also benefits, and in some cases dramatically. One film in Canada produced nearly $800,000 in taxes (*The Economist*, 1998). The direct economic impacts of film production are unquestionably large (Rich, 1997; Riley, Baker, and Van Doren, 1998), and research has shown indirect benefits to be positive with a multiplier effect of between 1.3 and 3.57 for film production, dependent upon the nature of the film (independent versus major production) and the site where filming takes place (Hydra Associates Limited, 1997; *The Economist*, 1998; Film New Zealand, 2000; Virginia Tourist Corporation, 2000). It is for this reason, and that of employment, that many regions now actively seek film production.

Film-induced Tourism

Film production, in addition to literary associations and television series, is also identified as a means to promote place and to reinvigorate image. Evidence suggests that movies can impact extensively upon on a location, not just during production, but also after the movie has been screened (Riley, Baker and Van Doren, 1998, Busby and Klug, 2001). The role of the feature film and its influences in tourism through post-production exposure (PPE) has been established (Riley and Van Doren, 1992, Riley, 1994, Riley, Baker and Van Doren, 1998, Busby and Klug, 2001). In tourism, the place of the feature film as a hallmark event, as a tourism-inducing event, and as tourism promotion has been recognised, and the effects of these on visitor numbers in predominately rural and natural areas have in some cases been immense, as exemplified in Table 9.3.

Table 9.3 by no means presents a complete list, and many other films are acknowledged to have led to increased visitation, albeit lacking the empirical support (Riley and Van Doren, 1992; Evje, 1998; Riley, Baker, and Van Doren, 1998; *The Economist*, 1998). The Australian films of *Crocodile Dundee* and *Crocodile Dundee II*, backed up by a number of other films created a surge in visitor arrivals to Australia (Riley and Van Doren, 1992; Hawker, 1998; New South Wales Film and Tourism Office, 1998). These have more recently been reinforced with the international promotion of *Babe* and *Mission Impossible 2* (New South Wales Film and Tourism Office, 1998; Tourism New South Wales, 2000). Thailand identified the benefits of film-induced tourism with the filming of *The Beach* (*Time*, 1999; Eaton, 2000; Tayman, 2000), and Hong Kong identified similar effects for *Rush Hour 2* (Asia Travel Tips, 2001).

Thailand has hosted other Hollywood productions that have led to an increase in tourism, such as *The Man With The Golden Gun*, *The King and I*, and its remake *Anna and the King* (actually filmed in Malaysia) (*The New Strait Times*, 1999; Eaton, 2000). The soft pornography film *Emmanuelle in Bangkok* also created a lasting impression of Thailand. In fact Bishop and Robinson (1998: 33) believed that it was 'almost impossible to write an article, whether focussed on politics or tourism, without at least glancing reference to what the reader already "knew" about Thailand from this single colourful source'.

The United Kingdom has been the screen location for many films and it is suggested that they have had a predominantly positive effect on tourism. Scotland's rugged scenery has been featured in many films (Gold and Gold, 1995). The many examples from Scotland and especially their effects on the American market are well known, at least anecdotally. This increase is attributed in part to the initiative the Scottish Tourist Board who 'persuaded MGM executives to run, for free, a travel advertisement before American showings of the film [*Rob Roy*]' (*The Economist*, 1995: 57). In addition other films such as *Braveheart*, *Loch Ness*, and *The Bruce*, have all created images of Scotland that have induced visitors (Stewart, 1997; System Three, 1997). For a few years in the 1990s, a significant part of Scotland's growth in the tourism industry and international tourist arrivals was attributed to these four films and was estimated to have introduced an additional £7-12mn in tourist expenditure (Hydra Associates Limited, 1997).

Other films such as *The Sound of Music*, have brought Salzburg (Austria) millions of dollars annually (Marriott, 2000). The effect is also evident in that Austria itself is known as the 'Sound of Music' country (Luger, 1992; Morgan and Pritchard, 1998). *The Commitments* is one Irish film to affect tourism, 'nurturing a fond image of the Irish and providing the kind of tourism promotion money just cannot buy' (Hegan, 1999: A15). Spain has also benefited from movie-induced tourism; a film location used for Western movies throughout the 1960s and 1970s has since been transformed into a tourist attraction and receives up to 2,000 visitors a day (Elphicke, 2001). The filming of *Hannibal* in Florence, Italy, initially concerned local officials due to the physical violence portrayed in the film and the effect that this may have had on the image of the city as well as any resulting impacts. Instead, the film site is offered for expensive, yet short, tours (Moldofsky, 2001). The incidence of film-induced tourism is also believed to be present in Africa where *Gorillas in the Mist* has had a positive impact for tourism, increasing tourism arrivals to Rwanda by 20 per cent in the year following its release (Hill, 1994). Another film that increased tourism visitation to Africa was *Out of Africa* (Tayman, 2000).

Table 9.3   Effects of rural and natural setting films on visitor numbers

Film	Location	Visitor Numbers
Brassed Off	Grimethorpe, England	50% increase in tourist numbers
Bridges of Madison County	Winterset, USA	An additional 220 coach parties, from 21 countries
Dances with Wolves	Fort Hayes, Kansas, USA	1990-91 – 25% increase, compared with 6.6% average increase over previous four years
Deliverance	Rayburn County, Georgia, USA	20,000 visitors
Field of Dreams	Dyersville, Iowa	55,000 visitors annually
Four Weddings and a Funeral	The Crown Hotel, Amersham, England	Fully booked for at least 3 years
Last of the Mohicans	Chimney Rock Park, North Carolina, USA	25% increase
Little Women	Orchard House, USA	65% increase
Mrs Brown	Osborne House, Isle of Wight	25% increase in tourist numbers
Notting Hill	Kenwood House, England	10% increase in tourists in one month
Remains of the Day	Durham Park, England	9% increase in tourist numbers
Saving Private Ryan	Normandy, France	40% increase in American tourists
Sense and Sensibility	Saltram House, England	An increase of tourist numbers by 39%
Steel Magnolias	Natchitoches, Los Angeles, USA	1990 – 39.7% increase
The First Knight	Trawsfynydd, Wales	15% increase in tourist numbers
The Fugitive	Great Smokey Mountain Railroad, USA	11% increase in train passengers
Thelma and Louise	Arches National Monument, Moab, Utah, USA	1988 – 11% increase 1989 – 8.8% increase 1990 – 3.3% increase

Sources: Riley and Van Doren (1992), Evje (1998), Riley, Baker and Van Doren (1998), and Busby and Klug (2001).

While most areas capitalise on their appearance in a film, some regions reinvent their own identity through films. For example, Riverside, Iowa USA, has promoted itself as the future birthplace of Captain Kirk (*The Economist*, 1999). The town has also erected a monument and hosts an annual *Star Trek* festival. Another US example is the use of *The Wizard of Oz* by Liberal, Kansas to reinvigorate its museum (attendance rose from 4,000 to 21,000), and develop new attractions (*The Economist*, 1999).

Tourism organisations have also identified the positive effects of films on tourism and have collaborated with the film industry in order to ensure that film production and PPE is maximised (Hydra Associates Limited, 1996, 1997). An example of this is the British Tourism Authority (BTA) attempting to secure more Indian film production for the benefits for tourism (Woodward, 2000). In attempting to further exploit the PPE of films the English Tourist Council (ETC) provides a film and site map for tourists. Maps were produced and distributed through travel agencies and the Internet to capitalise on popular films (Wintour, 1999). The ETC's implementation of the movie map has since culminated in the adoption of this method by the BTA for the whole of the UK. The map now features over 60 film locations and over 70 films (BTA, 2000). Independently, Manchester (northern England) has created its own coach tour of the sites in the city that have been used for screen locations (Schofield, 1996).

In summary, literary, television and film tourism are increasingly identified as motivating factors for tourists. Disenfranchised rural areas have also especially focused on this as a means of diversification of traditional economic development strategies and re-imaging their region. With many of these peripheral rural areas not having an association with literary arts, some are actively promoting their location to be part of a modern form of fictional media tourism. More recently they are promoting themselves as film production sites. Similarly, central and local governments are focusing on the economic significance of the feature film industry especially for recession-afflicted rural areas.

Film and Film Tourism in New Zealand

The production and the possible PPE impacts of feature films are being identified in New Zealand, especially after the large US$350mn budget trilogy of *The Lord of the Rings* was filmed in the country. Nonetheless, the first film to create tourism awareness of New Zealand was the hang-gliding and extreme skiing film *Off the Edge*, creating much exposure for the adventure tourism industry (Bruce, 2001).

Prior to the emphasis on *The Lord of the Rings* trilogy, *The Piano*, a critically acclaimed film by Jane Campion, was identified as a possible tourism generator (Hill, 1994), although support for this was not universal, with many arguing that the film was too arty and the scenery too gloomy (Hill, 1994; NZPA, 1994). Nonetheless, this was countered by arguments that *The Piano* would be positive for tourism and an opportunity for further promotion, and as good as *Crocodile Dundee* was for Australia in the preceding decade. *The Piano* was positive for tourism, and the Waitakere area (Auckland), where it was shot, has also achieved an international profile as a film production area, exemplified by being the site of the *Xena* and *Hercules* television series (Auckland Regional Council, 2000). Karekare Beach, where the piano was left in the film, is now a tourism site receiving visitors from around the world (Thompson, 2000). Reinforcing the images produced by the film, Tourism New Zealand, the national tourism organisation, used an image from *The Piano* in its *100% Pure New Zealand* campaign (Figure 9.1).

Source: Caroline Jackson. Marketing Communications Co-ordinator. Tourism New Zealand, 2003

Figure 9.1   *The Piano*, used in a New Zealand promotional poster

Whilst New Zealand is not identified as a film production location of scale, in recent years it has been utilised more frequently. Consequently, the economic significance of the film industry in New Zealand has increased and was reputedly worth NZ$500mn in the 1999-2000 financial year (SPADA, 2001).

The majority of this income was from large Hollywood productions located in New Zealand. *Vertical Limit* and *The Lord of the Rings* trilogy are the two most well known American big budget films produced in New Zealand, recently joined by *The Last Samurai*. Previously, other smaller international productions such as *Willow, The Rescue, White Fang* and *The Frighteners*, as well as a number of made for television series, movies, commercials, and documentaries have also been filmed in New Zealand. These international productions are an indication of a rapidly growing New Zealand film industry (Film New Zealand, 2000), additionally producing many small budget New Zealand focused and financed films, such as *Sleeping Dogs, Once Were Warriors, The Price of Milk*, and *Whale Rider*.

Although unknown to many people, the Indian film industry has also produced a number of films in New Zealand (Film New Zealand, 2001). These films are held to have an impact on the Indian tourism market imaging of New Zealand as a 'status' destination, exerting more impact than a marketing campaign (Williamson, 1999; Coventry, 2001; Hotton, 2001; Gupta, 2002). In fact, there was a 39 per cent increase in Indian visitors to New Zealand in 1999, and it was largely attributed to Kamal Singh, an Indian film production manager based in Christchurch (Cuzens, 2000).

Identifying these economic and PPE impacts, many New Zealand local authorities, which are traditionally rural based and primary industry dependent, have created film offices or adopted policies to further promote their areas and diversify their economic base. In conjunction with the New Zealand Film Commission, which facilitates the production of films in New Zealand by New Zealanders, local territorial authorities, regional tourism organisations and local film offices are assisting film production companies. These local offices have also been actively involved with the attraction of New Zealand and international film producers to their regions.

The increasing emphasis on attracting film production is largely based on perceived direct economic impacts, although the PPE effects reinvigorating the New Zealand image as a tourist destination have also been identified. Nonetheless, the data to support the positive effects of film on tourism arrivals in a New Zealand context are largely anecdotal, and no research has established this nor identified the impacts film productions have and can create. Nor have the authorities obtained information concerning the possible impacts and strategies required to maximise the image benefits of being a film location.

To rectify this predicament and within the context of increasing emphasis on film production, all of New Zealand's local government offices and regional tourism organisations (RTOs) were surveyed by the authors in 2001 with a mail-out/mail-back survey (a representative 54 per cent responded). Three events were felt to facilitate the implementation of the survey instrument at this time. First, there was increasing publicity about *The Lord of the Rings* trilogy being filmed around the country at the time (Maling and Espiner, 2000; Campbell, 2001; McGrath and Beatson, 2001). Second, there was increasing publicity of the dramatic benefits associated with filming in the media, and especially local government's involvement in facilitating this (Film New Zealand, 2000; Haines, 2000; Queenstown-Lakes District Council, 2000; Campbell, 2001). Third, central government had implemented changes to its regional development policy emphasising localised initiatives (Ministry of Economic Development, 2001).

The research identified the role New Zealand local government gave to film production, especially as a means to develop local areas' economy and tourism. It was particularly focused on eliciting local government objectives for the development of the film industry, the methods that were being considered to maximise the potential promotional and touristic benefits of the film industry, and finally considerations for the management of image.

Most of the regions of New Zealand had been used for film production of some kind, reflecting a geographical dispersal and absence of a core production area. Film crews had come from the USA, India, the United Kingdom, Canada, Japan, Singapore, China and other Asian countries, France and other continental European countries.

From these productions only 17 per cent of the regions had noticed increases in tourism, highlighting the new nature of the film industry and the haphazard way that its effects are being monitored. Feature films were noted as the greatest tourism-inducing event. Nonetheless television documentaries, the *Xena: Warrior Princess* and *Hercules* television series, and advertisements, and in particular the *L&P* (New Zealand soft drink brand) promotion were noted to also induce tourism. Even with the low number of regions experiencing increased tourism, 71 per cent thought that tourism would increase. Due to this, and more importantly to the economic and employment impacts of film production, almost half of all New Zealand regions have included film within their economic development strategies. Other stated reasons for placing importance on film production for the community include image enhancement and cultural development.

Even though image enhancement is the main creator of film-induced tourism, most New Zealand regions did not place importance on being selective in the choice of films produced in their area, focusing more on the short-term economic impacts of film production. A film's merit needs to be assessed in terms of its promotional value, as films can increase awareness in a positive manner and therefore move the screened destination into a choice set of destinations (Woodside and Sherrell, 1977; Moutinho 1987, Um and Crompton, 1990, Goodall, 1991). Film-induced tourism has occurred across a variety of images and thematic contents, although image and theme do not seem to have been determinates. In addition, as noted in the *Hannibal* case, images perceived as detracting from a location may not negatively affect the holistic image that viewers obtain. In addition to the economic objectives in the attraction of film production, many New Zealand local government offices acknowledged that they were not legally entitled select production genres.

Even without that selectivity, two thirds thought that films produced in their area could be used as promotion. Three conclusive categories of respondents were identified in responses to using films for promotion. The premise of the first group was that exposure is positive, reflecting a common notion that any publicity is good publicity. The second group concentrated more on promotional benefits in terms of future film production. The third group thought that films could be used as promotion but it would depend on factors such as what type of film, what was being shown, the location that was being used and whether the region would be known by the viewer. This reflects Riley and Van Doren's (1992) suggestion that the viewer's knowledge of the location's identity was a limiting factor for film-induced tourism.

Retention of local image was deemed important for respondents, as was having an internet site linking film productions to the region. Internet linking of film to place has recently been further emphasised in New Zealand where Tourism New Zealand has developed part of its website specifically promoting *The Lord of the Rings* and its film sites throughout New Zealand (*New Zealand Home of Middle-Earth*, <http://www.purenz.com/homeofmiddleearth>). There was also increased emphasis on using films' stars to promote the local area to tourists.

Overall, the research showed that respondents tended to look favourably upon the film industry with those that did not have it in their area being keen to develop the industry. The film industry is sought for several reasons and although the potential for film production to increase tourism was widely acknowledged, it was primarily the economic benefits that were sought. Although some respondents did recognise the potential benefits of feature film production to affect the tourism industry in their area, it appeared that little planning had been conducted to maximise these impacts, and that the PPE impacts on image were not fully understood. Respondents exemplified this in that many lacked the specific knowledge as to how their organisation could participate in the development of the film industry. While areas actively seek film production it would appear that research has to be undertaken by these organisations to gain an understanding of the processes involved. If this knowledge was gained then it is possible that the production and PPE impacts could be better controlled in the interests of the community, image and tourism stakeholders. To this end regional film clusters, Trade New Zealand, and Film New Zealand are making progress (New Zealand Institute of Economic Research, 2002). The impacts of film-induced tourism still appear under-appreciated even though they can be long lasting and have significant long-term economic and social effects (Riley and Van Doren, 1992).

Conclusion

A decline in the economic base of many rural areas through forces of globalisation and other factors emphasises the need for susceptible areas to diversify. Tourism has been identified as a means to do this, and in rural areas it has been found that associations with the fictional media can be utilised. Nonetheless many rural locations do not have such associations and therefore in efforts to develop similar linkages, some promote their areas to be television or film settings. The use of rural areas as film settings creates an initial economic impact, though more importantly for the further development of tourism and destination management are the PPE impacts, such as image reinvigoration and film-induced tourism.

For destination management, the image portrayed, and perceived by potential visitors, is very important, and therefore, should be given special attention in destination management planning. The evaluative components of destination image as displayed in the media are critical for destination management and marketing, especially as it has been identified that film can act as promotion of the destination, promotion for possible visitors and production companies. The possible impacts of image projection in film-induced tourism have been identified as potentially complementary to the objectives of economic development.

Films have been identified as a tourist-inducing element, turning places of no note into much visited tourist attractions. Although not all films have such an effect, nonetheless, it can be argued that film, and the media in general do affect the image of place. It is therefore important that a considered image management plan be created for destinations. This is vital to ensuring the sustainable and successful development of the destination. In the development of the image management strategy a long-term view is needed. The results of research in New Zealand show a short-term focus that facilitates film production, concentrating on the associated economic impacts. However, an appreciation of film as a positive image enhancer is also needed. Planning needs to adopt a long-term consideration of image and the PPE impacts of film on this image.

Within image management strategic plans, an accurate assessment of the evaluative components of the destination and the components that visitors may evaluate should be undertaken. Mitigation or enhancement strategies should be developed within the image management strategy for use when opportunities or threats arise. These images can be produced by many events, including hallmark events, disasters, and terrorism. Even though many destination management organisations cannot be selective of films being produced, they can be proactive in promoting their location to film producers. Consideration should be given as to the best film genres for disseminating the proposed image to visitor groups. The promotional ability of films cannot be assumed to be consistent, as highlighted by the low-key nature of much of the New Zealand film industry. Nonetheless, when the opportunity arises this needs to be recognised and allocated appropriate resources.

Overall, this chapter has shown that the fictional media can be a source of tourism promotion and one that many RTOs and local governments in New Zealand have yet to appreciate fully. Although the economic development opportunities presented by film production are being consciously exploited, it is the PPE effects on image and the effects of film-induced tourism that need to be understood and appreciated. Recognition of these opportunities needs to be developed within these organisations in order to maximise the economic development potential of film production, particularly for rural areas needing to diversify their economic base.

Finally, there are many areas requiring further research within the relationship of destination management, economic development, film and tourism. The study of film is relatively new in tourism research, and more research is needed at the destination level to assess the evaluative components of image and to measure the effect the media, and in particular films, have on image.

References

A C Nielson (1999), *Cinema Industry*, A C Nielson, Wellington.
Aitken, S.C. and Zonn, L.E. (eds) (1994a), *Place, Power Situation and Spectacle: A Geography of Film*, Rowman and Littlefield, Maryland.

Aitken, S.C. and Zonn, L.E. (1994b), 'Re-Presenting the Place Pastiche', in S.C. Aitken and L.E. Zonn (eds), *Place, Power Situation and Spectacle: A Geography of Film*, Rowman and Littlefield, Maryland, pp. 3-25.

Altheide, D.L. (1997), 'Media Participation in Everyday Life', *Leisure Sciences*, vol. 19, pp. 17-29.

Anderson, K. and Gale, F. (eds) (1992), *Inventing Places: Studies in Cultural Geography*, Longman, Melbourne.

Asia Travel Tips (2001), 'Latest Travel News: Hong Kong Stars in Rush Hour 2 Debut', *Asia Travel Tips.com* <http://www.asiatraveltips.com/30July2001HKTB.htm>.

Auckland Regional Council (2000), *Auckland Regional Council: Waitakere Ranges – Regional Park*, Auckland Regional Council, Auckland.

Australian Film Commission (2003), 'Numbers of Australian Cinema Admissions and Gross Box Office', 1976-2001, *Australian Film Commission* <http://www.afc.gov.au/resources/online/gtp_online/Cinema/Admission_numbers_table.htm>.

Beeton, S. (2000), '"It's a Wrap!" But What Happens After the Film Crew Leaves?' *Travel Tourism Research Association Annual Conference Proceedings*, California, 11-14 June.

Beeton, S. (2001a), 'Lights, Camera Re-Action: How Does Film-induced Tourism Affect a Country Town?', in M. Rogers and Y. Collins (eds), *The Future of Australia's Country Towns, Australia. Centre for Sustainable Regional Communities*, La Trobe University, Bendigo, pp. 172-83.

Beeton, S. (2001b), 'Smiling for the Camera: The Influence of Film Audiences on a Budget Tourism Destination', *Tourism, Culture and Communication*, vol. 3, pp. 15-26.

Belch, G.E. and Belch, M.A. (1998), *Advertising and Promotion: An Integrated Marketing Communications Perspective: International Edition*, McGraw Hill, Boston, 4th edition.

Bishop, R. and Robinson, L.S. (1998), *Night Market: Sexual Cultures and the Thai Economic Miracle*, Routledge, New York.

British Film Institute (2003), 'UK Cinema Admissions 1933-2001', *British Film Institute* <http://www.bfi.org.uk/facts/stats/alltime/uk_admissions.html>.

British Tourism Authority (BTA) (2000), 'Movie Map Online', *BTA*, London <http://www.visitbritain.com.moviemap/moviemap/moviemap.html/>.

Bruce, D. (2001), 'NZ Hang-Gliding Pioneer Returns', *Otago Daily Times*, 18 January, p. 11.

Busby, G. and Klug, J. (2001), 'Movie-Induced Tourism: the Challenge of Measurement and Other Issues', *Journal of Vacation Marketing*, vol. 7, pp. 316-32.

Butler, R.W. (1990), 'The Influence of the Media in Shaping International Tourist Patterns', *Tourism Recreation Research* , vol. 15, pp. 46-53.

Butler, R.W., Hall, C.M. and Jenkins, J.M. (eds) (1998), *Tourism and Recreation in Rural Areas*, John Wiley & Sons, Chichester.

Campbell, G. (2001), 'Absolutely, Positively Middle Earth', *Listener*, 3 February, pp. 33-4.

Cater, E. (2001), 'The Space of the Dream: A Case of Mis-Taken Identity?' *Area*, vol. 33 (1), pp. 47-54.

Coates, J.F. (1991), 'Tourism and the Environment: Realities in the 1990s', *World Travel and Tourism Review: Indicators, Trends and Forecasts*, vol. 1, pp. 66-71.

Coventry, N. (2001), 'Film Influence is Big', *Inside Tourism*, p. 345.

Curtis, James R. (1981), 'The Boutiquing of Cannery Row', *Landscape*, vol. 25, pp. 44-8.

Curtis, J.R. (1985), 'The Most Famous Fence in the World: Fact and Fiction in Mark Twain's Hannibal', *Landscape*, vol. 28, pp. 8-14.

Cushman, G., Veal, A.J. and Zuzanek, J. (eds) (1996), *Leisure Participation: Free Time in the Global Village*, CAB International, Wallingford.

Cuzens, S. (2000), 'Indian Film Makers Bring Tourists to New Zealand', *Business Development News*, (76), p. 5.

Deiss, R. (2001), *Statistics in Focus: Cinema Statistics: Strong Growth in Cinema-going*, Eurostat, Luxembourg.

Eaton, Dan (2000), 'Thai Tourism Banks on Boost', *Otago Daily Times*, 24 February.

Elphicke, C. (2001), 'For a Few Dollars More', *Air New Zealand Panorama*, February, pp. 65-6.

Evje, B. (1998), 'Baseball Heaven: Iowa's Mythical Field of Dreams is Where Reality Meets Fantasy', *Sport*, vol. 89, p. 45.

Fawcett, C. and Cormack, P. (2001), 'Guarding Authenticity at Literary Tourism Sites', *Annals of Tourism Research*, vol. 28, pp. 686-704.

Film New Zealand (2000), Film NZ Strategic Plan, *Film New Zealand*, Wellington.

Film New Zealand (2001), 'India: No Lesser Dreams', *FilmNZnews*, vol. 1, p. 15.

Fitzgerald, M. (2000), 'Harbouring Hollywood', *Time*, 31 July, pp. 32-6.

Fodness, D. and Murray, B. (1999), 'A Model of Tourist Information Search Behaviour', *Journal of Travel Research*, vol. 37, pp. 220-30.

Gartner, W.C. (1993), 'Image Formation Process', *Journal of Travel and Tourism Marketing*, vol. 2, pp. 191-215.

Gold, J.R. and Gold, M.M. (1995), *Imagining Scotland: Tradition, Representation and Promotion in Scottish Tourism Since 1750*, Scolar Press, Aldershot.

Goodall, B. (1991), 'Understanding Holiday Choice', *Progress in Tourism Hospitality Research*, vol. 3, pp. 58-77.

Gupta, K. (2002), 'Yes, We Want Them, but How Do We Get Them: India as an Emerging Market for Tourism New Zealand', in W.G. Croy (ed.), *International Tourism Students Conference Proceedings*, School of Tourism and Hospitality, Waiariki Institute of Technology, Rotorua, pp. 7-14.

Haines, L. (2000), 'Cast of Five Eyes Film NZ's Role', *The Dominion*, 25 March, p. 10.

Hall, C.M. and Kearsley, G. (2001), *Tourism in New Zealand: An Introduction*, Oxford University Press, Melbourne.

Hanna, S.P. (1996), 'Is it Roslyn or is it Cicely? Representation and the Ambiguity of Place', *Urban Geography*, vol. 17, pp. 633-49.

Hawker, W. (1998), 'There's Room at the Top', *Export News*, vol. 32, p. 54.

Hegan, C. (1999), 'Long-term Profit in Quotas', *New Zealand Herald*, 7 July, p. A15.

Herbert, D.T. (1996), 'Artistic and Literary Places in France as Tourist Attractions', *Tourism Management*, vol. 17, pp. 77-85.

Herbert, D. (2001), 'Literary Places, Tourism and the Heritage Experience', *Annals of Tourism Research*, vol. 28, pp. 312-33.

Hill, D. (1994), 'Beatson Doesn't Play The Piano', *Admark*, 31 March, p. 48.

Hotton, M. (2001), 'Indian Film Set in Larnach Castle', *Otago Daily Times*, 10 February, p. 1.

Hunter, J. (1999), 'Northern Exposure: More TV Shows and Movies are Being Filmed in Canada, Despite Hollywood's Complaints', *MacLean's*, 11 October, p. 68.

Hydra Associates Limited (1996), *A Comparison of the Factors Affecting Feature Film Production in Six Countries*, Report prepared for British Screen Advisory Council, British Film Commission, Ernst & Young and Producers Alliance for Cinema and Television, London.

Hydra Associates Limited (1997), *The Economic and Tourism Benefits of Large-scale Film Production in the United Kingdom*, Report prepared for British Film Commission, Scottish Screen Locations, Scottish Tourist Board, Edinburgh.

International Trade Administration (2001), *The Migration of U.S. Film and Television Production: Impact of 'Runaways' on Workers and Small Business in the U.S. Film Industry*, United States Department of Commerce, Washington D.C.

Jackson, P. (1992), 'Constructions of Culture, Representations of Race: Edward Curtis's "Way of Seeing"', in K. Anderson and F. Gale (eds), *Inventing Places: Studies in Cultural Geography*, Longman, Melbourne, pp. 89-106.

Kotler, P., Haider, D.H. and Rein, I. (1993), *Marketing Places: Attracting Investment, Industry, and Tourism to Cities, States, and Nations*, Macmillan, New York.

Luger, K. (1992), 'The "Sound of Music" Country: Austria's Cultural Identity', *Media, Culture and Society* , vol. 14, pp. 185-93.

Maling, N. and Espiner, G. (2000), 'Curtain Falling on Ring Master's Circus', *Sunday Star Times*, 12 November, p. 23.

Marriott, B. (2000), '"Sound of Music" City Satisfies', *Otago Daily Times*, 14 November, p. 20.

McGrath, J. and Beatson, P. (2001), 'Cannes Critics Spellbound by Rings', *Sunday Star Times*, 13 May, p. A1.

Ministry of Economic Development (2001), 'Enhancing Economic Transformation and Growth. Industry and Regional Development', *Ministry of Economic Development*, Wellington <http://www.med.govt.nz/irdev/asst_prog/budget2001/enhancing.html>.

Moldolfsky, L. (2001), 'Traveller's Advisory: Europe – Florence', *Time*, 19 March, p. 4.

Monitor Company (2001), 'US Runaway Film and Television Production Study Report', *Film and Television Action Committee* <http://www.ftac.net/dga_sag_report/ filmreport.htm>.

Morgan, N. and Pritchard, A. (1998), *Tourism Promotion and Power: Creating Images, Creating Identities*, John Wiley & Sons, Chichester.

Motion Picture Association (2002), '2001 US Economic Review: Box Office', *Motion Picture Association of America* <http://www.mpaa.org/useconomicreview/ 2001Economic/sld005.htm>.

Moutinho, L. (1987), 'Consumer Behaviour in Tourism', *European Journal of Marketing*, vol. 21, pp. 3-44.

Muller, D.K. (2001), 'Literally Unplanned Literary Tourism in two Municipalities in Rural Sweden', in M. Mitchell and I. Kirkpatrick (eds), *New Directions in Managing Rural Tourism and Leisure: Local Impacts, Global Trends*, SAC, Auchincruive, CD-Rom.

Muresan, A. and Smith, K.A. (1998), 'Dracula's Castle in Transylvania: Conflicting Heritage Marketing Strategies', *International Journal of Heritage Studies*, vol. 4, pp. 73-85.

New South Wales Film and Tourism Office (1998), 'Everything You Need to Know About Developing Model Film Policy', *New South Wales Film and Tourism Office* <http://www.ftosyd.nsw.gov.au/TEXTONLY/ftohome.htm>.

New Zealand Film Commission (2000), 'About Us', *New Zealand Film Commission* <http://www.nzfilm.co.nz/frame4aboutus.html>.

New Zealand Institute of Economic Research (2002), *Scoping the Lasting Effects of The Lord of the Rings: Report to the New Zealand Film Commission*, New Zealand Institute of Economic Research, Wellington.

Nielsen, C. (2001), *Tourism and the Media: Tourist Decision-Making, Information, and Communication*, Hospitality Press, Elsternwick.

NZPA (New Zealand Press Association) (1994), '"Piano" Gives NZ World Springboard', *New Zealand Herald*, 24 March, p. 24.

O'Connor, N. (2001), 'The Effect of Television Induced Tourism on the Village of Avoca, County Wicklow', *ATLAS*, October. pp. 1-22.

Queenstown-Lakes District Council (2000), *Queenstown Film Friendly*, Queenstown-Lakes District Council, Queenstown.

Rich, B.R. (1997), 'Brazilian State Bids for Piece of Film Biz Pie', *Variety*, vol. 367, p. 23.

Riley, R.W. (1994), 'Movie Induced Tourism', in A.V. Seaton, C.L. Jenkins, P.U.C. Dieke, M.M. Bennet, L.R. MacLelland and R. Smith (eds), *Tourism: The State of the Art*, John Wiley & Sons, Chichester, pp. 453-8.

Riley, R.W. and Van Doren, C.S. (1992), 'Movies as Tourism Promotion: A "Pull" Factor in A "Push" Location', *Tourism Management*, vol. 13, pp. 267-74.

Riley, R.W., Baker, D. and Van Doren, C.S. (1998), 'Movie Induced Tourism', *Annals of Tourism Research*, vol. 25, pp. 919-35.

Roberts, L. and Hall, D. (2001), *Rural Tourism and Recreation: Principles to Practice*, CAB International, Wallingford.

Schofield, P. (1996), 'Cinematographic Images of a City: Alternative Heritage Tourism in Manchester', *Tourism Management*, vol. 17, pp. 333-40.

Seaton, A.V. (1999), 'Book Towns as Tourism Developments in Peripheral Areas', *International Journal of Tourism Research*, vol. 1, pp. 389-99.

Small, K. and Edwards, D. (2002), 'Evaluating the Socio-Cultural Impacts of a Festival on a Host Community: A Case of the Australian Festival of the Book', in W.G. Croy (ed.), *New Zealand Tourism and Hospitality Research Conference Proceedings*, School of Tourism and Hospitality, Waiariki Institute of Technology, Rotorua, pp. 150-67.

SPADA (Screen Producers and Directors Association of New Zealand) (2001), *Survey of Screen Production in New Zealand 2000*, Screen Producers and Directors Association of New Zealand, Wellington.

Squire, S.J. (1994), 'The Cultural Values of Literary Tourism', *Annals of Tourism Research*, vol. 21, pp. 103-20.

Squire, S.J. (1996), 'Literary Tourism and Sustainable Tourism: Promoting "Anne of Green Gables" in Prince Edward Island', *Journal of Sustainable Tourism*, vol. 4, pp. 119-34.

Stewart, M. (1997), 'The Impacts of Films in the Stirling Area: A Report. Scottish Tourist Board', *Research Newsletter*, 12 July, pp. 60-6.

System Three (1997), *Impacts of Films Study: Final Report*, System Three for Stirling Council Economic Department, Stirling.

Tayman, J. (2000), 'Footsteps on the Beach: Trouble on Fantasy Island', *Outside Magazine*, January <http://www.thaistudents.com/thebeach/outside.html>.

The Economist (1995), 'Rob Roy to the Rescue', *The Economist*, 13 May.

The Economist (1998), 'United States: Lures and Enticements', *The Economist*, vol. 346 (8059), pp. 28-29.

The Economist (1999), 'Moreover: Where Bluebirds Fly', *The Economist*, vol. 353 (8143), pp. 97-98.

The New Strait Times (1999), '"Anna and the King" To Boost Tourism in Perak', *The New Strait Times*, 2 February, p. 7.

The Numbers (2003), 'All Time Top 20 Movies by Global Box Office', *The Numbers: Box Office Data, Movie Stars, Idle Speculation* <http://www.the-numbers.com/movies/records/index.html#world>.

Thompson, S. (1992), 'Lights! Camera! (Economic) Action!', *American City & County*, vol. 107, pp. 28-35.

Thompson, W. (2000), 'Disney Heads to Henderson', *New Zealand Herald*, 19 August, p. 10.

Time (1999), 'In the Swim Again', *Time*, vol. 153, p. 64.

Tooke, N. and Baker, M. (1996), 'Seeing is Believing: the Effect of Film on Visitor Numbers to Screened Locations', *Tourism Management*, vol. 17, pp. 87-94.

Tourism New South Wales (2000), 'Tourism New South Wales Annual Report 1999-2000', *Tourism New South Wales* <http://www.tourism.nsw.gov.au/corporate/downloads/Annualreport.pdf>.

Townson, D. (2000), 'Hollywood Productions Head West', *Variety*, vol. 378, p. 158.

Um, S. and Crompton, J.L. (1990), 'Attitude Determinants in Tourism Destination Choice', *Annals of Tourism Research*, vol. 17, pp. 432-48.

United Nations (2000), 'Is the United Nations a Good Investment? Image and Reality', *United Nations*, New York <http://www.un.org/geninfo/ir/ch6/ch6.htm>.

Virginia Tourist Corporation (2000), 'Virginia Film: Past, Present and Future', *Virginia Tourist Corporation* <http://www.vatc.org/film/history>.

Weir, D.T.H. (2003), 'The Literary Heritage of Neville Shute', in M. Robinson and H.C. Andersen (eds), *Literature and Tourism: Essays in the Reading and Writing of Tourism*, Continuum, London.

Weller, J. (1999), 'Testimony Before the Subcommittee on Trade of the House Committee on Ways and Means', 5th August, Hearing on the United States Negotiating Objectives for the WTO Seattle Ministerial Meeting <http://waysandmeans.house.gov/legacy.asp?file=legacy/trade/106cong/tr-3wit.htm>.

Williamson, K. (1999), 'One Billion Will Watch Film', *Otago Daily Times*, 20 August, p. 9.

Wilson, C. (1996), *Landscapes in the Living Room: Heartland Documentaries and the Construction of Place*, unpublished Master of Social Science thesis, University of Waikato.

Wintour, P. (1999), 'Film Sites Used to Map Out Tourism Plan', *Otago Daily Times*, 1 July.

Woodside, A.G. and Sherrell, D. (1977), 'Traveller Evoked, Inept, and Inert Sets of Vacation Destinations', *Journal of Travel Research*, vol. 16, pp. 14-18.

Woodward, I. (2000), 'Why Should The UK's Tourism Industry be Interested in "Bollywood" Films?' *Tourism Intelligence Papers*, British Tourism Authority, London.

PART 4
STRATEGY AND MANAGEMENT

Chapter 10

Strategy Formulation in Rural Tourism – An Integrated Approach

Hans Embacher

Introduction

Tourism and the leisure industry are vitally important to the Austrian economy and play a key role in economic growth, employment and the balance of goods and services. With foreign exchange earnings of more than €14.5bn in 2000, tourism is unquestionably one of the cornerstones of the Austrian economy, and where such earnings from tourism represent approximately €1.74 per head of population (1999 figures), Austria can be seen to have by far the most intensive tourism amongst the Western industrialised nations.

Within the agricultural sector there is a strong need to generate additional income, ideally from resources on the farm (Ilbery and Bowler, 1998: 75). Therefore farm holidays are an ideal activity to use the available resources of a farm economically. From the total Austrian tourist supply, some 15,500 farmers offer 170,000 tourist beds as rooms or apartments. This means that approximately 8 per cent of all Austrian farmers offer tourist accommodation, representing one-fifth of all tourist enterprises and one-seventh of the total Austrian supply of tourist beds. Of all farmers involved in tourism about two-thirds are farmers in mountainous regions. Farm holidays are therefore an important economic segment in agriculture and tourism, particularly in the economies of mountain regions.

In 1988 the Austrian Agricultural Ministry formulated the 'eco-social way' for Austrian agriculture within which, in 1989, the first strategy for the farm holiday sector was developed. Since then, some 3,400 tourist farms (with 44,800 bed spaces) have become members of the Austrian Farm Holidays Organisation (about one quarter of all providers).

This chapter explores the conditions under which a new strategy for the farm holiday sector has emerged, and describes the processes involved in the new strategy's formulation based on the results of ten years of experiences within the Austrian Farm Holidays Association. It was necessary – with the involvement of a number of stakeholder groups – to find ways to become more flexible and to organise resources to meet the challenges of the future.

The results clearly show that the new strategy has successfully passed through its 'pioneer' phase and is now in the process of challenging the familiar in the creation of a new, more professional and coherent body to replace the nine separate (eight Provincial and one Federal Association) organisations that have hitherto existed.

This change will be an ongoing process and presents major challenges for everyone involved in the integration of agriculture and tourism in Austria.

Why a New Strategy?

The original strategy had been worked out under the assumption that future trends and developments are more or less predictable. In support of early strategic plans an organisational structure had been built up on several levels (Federal, Provincial and Regional). However, despite successful work and innovative projects the Organisation became increasingly 'tied down' with continuous routine work. It became evident that the increasing speed and pressure of change in the market and the consequent need for continuous innovation were beginning to threaten the early successes of the Organisation and consequently of the farm tourism sector and that a new strategy would be required to support tourism providers in the future.

At the beginning of the new strategy development process there were several challenges:

- to evaluate the development of the farm tourism sector and the work performed by the organisation in the past decade;
- to assess the effectiveness (are the right things done?) and efficiency (are the things done in the right way?) of the organisation in relation to the wide network of partners in product development, education/training and marketing (e.g. chambers of agriculture, Ministries, National and Provincial tourism organisations);
- to assess market opportunities;
- to involve practitioners, especially farmers, in the strategy formulation process in order 'to ground' the strategy;
- to make sure that a great deal of the strategy would be translated into reality by the various stakeholders concerned; and
- to formulate a strategy for the future considering the need for continuous change.

In order to prepare the Organisation for future challenges it was important to ask two important questions of key target groups. Of landlords: is our organisation working well? Of the public bodies in charge of granting subsidies: has the public money been invested in an effective way?

The Process

The development of the strategy was designed and executed as a process to encourage high stakeholder involvement in order to avoid producing an 'academic' paper that did not reflect everyday experience. The process was guided and accompanied by two companies: one tourism consultancy and a consultants' network with expertise in organisation development. The phases of the process were as follows:

Initial Workshop

At the first two-day-meeting with representatives of all stakeholder groups (landlords, staff, ministry, chamber of agriculture) objectives were specified, the process explained, and a timetable set. A steering group was established for the purposes of taking current decisions and keeping contact with the consultants.

Analysis

An analysis of major tourism trends, of supply and demand for farm and rural holidays in major European markets was carried out in the first phase. There were also telephone interviews with a selection of members. Through an 'ABC' rating a representative sample of members was sought. In the nine organisations at Federal and Provincial levels, the consultants carried out an organisational check to assess strengths, weaknesses and respective challenges for each association (with consideration given to their inter-dependence) and of the organisation as a whole.

'Future' Conference

After about six months' work, in a 2½ day workshop with some 25 participants, (from the organisation on each level and from major partners on Federal level), the following had been achieved:

- a visionary goal for 2010;
- the core strategies required to achieve the visionary goal;
- the leading principles along which the farm holidays product and the supporting organisation should be developed; and
- the means by which to implement the strategies as everyday business practices.

The results of this workshop provided the guidelines for subsequent developments.

Provincial Conferences

The results of analysis were presented to stakeholders and partners in each of the eight provinces and discussed with them in a half-day workshop. The objectives were to communicate the contents of the process to a wider target group, to hear their opinions and to integrate the perspective of the practitioners at both provincial and regional levels. Attendance at these provincial conferences ranged from groups of 20 to groups of 60 persons.

Partner Conference

In a partner conference with participants from the National Tourism Organisation, provincial tourism organisations, Austrian television, the Austrian Agricultural Marketing Organisation, rural development organisations, and business partners, the situation of the farm holidays product in general, and of the Association in particular, was assessed and discussed. The partners participated in discussion regarding the directions farm holidays should take in the future.

Culture Workshop

As there had been a lot of pressure thus far 'to produce results', there was an urgent need to integrate 'soft elements' like personal perceptions, assumptions, fears, relationships, criticism, appreciation of the performed work, for example, into development processes. These factors were analysed and discussed in a 2-day workshop after a good year's work. As the work in the organisation is characterised by a very high personal involvement this was a vital part of the process to clear out potential difficulties and obstacles at a personal level.

Project Management Workshop

Although much of the work so far had been carried out in the form of projects, members of the organisation were not familiar with the systematic approach of project management. In this workshop, key players in Federal and Provincial organisations were made familiar with issues such as: defining a project; project clients; project progress and supervision; communication of, and reporting on, process and results; and delivery of the results to the relevant bodies of the organisation that have to take decisions.

Parallel Operational Projects

At the '*Future*' conference, the first operational project had been started and several others soon followed. The projects were led either by the Federal Association or by a representative of one of the Provincial Associations. The idea was to start the implementation projects whilst the strategy process was still running and the groups as well as the expertise of the consultants still available.

The issues of the individual projects were: organisation development; membership strategy; development of enterprises (holiday farms), education/training; members' information including a magazine; marketing; branding; lobbying; *'Future'* workshop; quality control; and membership research.

As a result, a wide range of topics was covered and, through the participation of the provinces in the running of nationwide projects, the acceptance of the results could be raised. However the projects also resulted in a heavy workload for those frequently involved in several projects at the same time. The invitation of outside partners to contribute to various projects proved to be very valuable, and also provided a wider perspective of the topic.

In retrospect, the timing of the new strategy formulation was ideal for a number of reasons. The organisation had gained a lot of experience and was (still) working well and successfully. Nevertheless there were symptoms of stress and necessary change, although at this time it was possible to finance such a large-scale and expensive process.

Results/Findings

In the following section some of the findings from the experts' analysis as the basis for necessary changes and for the future strategy are outlined.

Members' Satisfaction

The representative survey showed a high rate of satisfaction with the organisation – fundamentally, all was well. An interesting result was that passive members (who did not attend meetings or usually did not return questionnaires) stated in a telephone survey that they were quite happy too – they simply did not attend meetings.

Efficiency of the Organisation/Co-operation

After in-depth analysis the experts stated that each association on Federal and Provincial level works efficiently with the given resources. To optimise the effectiveness (are the right things done?) the lack of an updated strategy and mid-term planning was criticised. It was also found that many tasks were carried out by each organisation in their own way instead of developing common procedures and by doing so benefiting from considerable synergies.

In some provinces the experts also pointed out the danger of being under-staffed and the resulting risk that the members' requirements and expectations could not be met due to the lack of resources.

Membership/Participation

Within the agricultural sector there is a strong need to generate additional income, ideally from resources on the farm. Therefore farm holidays are an ideal activity to use the available resources of a farm economically. The members of the organisation on average achieve a higher price (approximately 25 per cent above the average price of the whole sector) and a better load factor (about one-third above the statistical average of all holiday farms) than non-members. Despite this, only about one quarter of all holiday farms are members of the Organisation. So it has been strongly recommended to widen the membership basis.

Organisation Development

The Austrian Farm Holidays Organisation had developed from the 'pioneer stage' via a 'differentiation phase' towards the so called 'integration phase' with all consequent effects and necessary changes. Some of the original procedures from the founding phases needed to be checked and revised.

Market Chances/Potential

Research showed that there is considerable market potential for the farm holiday-product (Deutsche Reiseanalyse, 2003): in Germany some 8.4mn adults state their interest in farm holidays. In Austria, an additional 3.2mn persons are generally interested in this form of vacation. However, from this potential only about one-quarter have definitely decided on spending their holidays on a farm, the remaining three-quarters have to be convinced by the use of all available promotion channels. The relatively low rate of exploitation of the existing potential is amongst other reasons an expression of the limited media budgets.

It has been recommended to concentrate on the German speaking regions of Europe, although France and Italy would also be markets with interesting potential. Experience shows that the highest demand potential is to be found in countries where attractive holiday farms are on offer on the home market.

Individuality

Solutions that have been successful so far on each level are to a high degree characterised by their individuality. This has the great advantage that structural resources of the association and the personal strength of the individual are used to an optimum. On the other hand many of the positive effects of joint and standardised solutions are not achieved.

Visionary Objectives

By the year 2010 'Farm Holidays in Austria' should be a well-known international brand in the field of rural tourism with high involvement of the members in the development and marketing of the product. The organisation should have 5,000 member enterprises that strongly represent and reinforce the contents and ideas of the common brand.

Changes/New Ways

It is not possible within this chapter to provide a detailed account of the new strategy. However, the most important changes resulting from the new strategy development process are now outlined:

Administration

The accounting-system has been adapted so that all organisations use the same accounting software and systems. Thus it is now possible to compare and consolidate the business figures and office procedures have been streamlined.

In Austria in 2002 a new law for associations has been introduced. To cope with the necessary changes it was necessary to work out one solution for all part-organisations and therefore save a lot of work for each Provincial office.

Membership Strategy

A set of measures has been formulated in order to gain more members for the association. At the same time it is to be expected that some of the existing members might leave due to higher membership contributions and stricter requirements in terms of quality. On average the membership fee has been raised by 20-30 per cent and changed from a fixed amount to a variable sum according to the number of tourist beds on the farm.

Before the new strategy was implemented the members of the provincial associations were obliged to 'buy the whole package' which included brochure entries and internet exposure. With the new strategy the basic membership only includes the quality control and allows the member to use the 'Holiday on the Farm' brand. Members can subsequently decide whether to buy one or more additional modules such as space in the brochure or on the internet, placement in a special-interest brochure, e.g. horse riding; whether to become member of a project group like 'seminars on the farm', sell the farm's alpine chalet through the organisation or to take part in a cross-border INTERREG-project.

The idea is that the members can choose exactly the amount of support they need for their given situation.

Quality Classification

The existing classification system (with the categories of two, three and four flowers) has been revised and a more integrated approach adopted. So far the emphasis of the criteria was on the services provided by the landlords (in order to ensure the visitor's holiday experience). Now the farm has to provide top quality in the three areas of 'farm quality', 'facilities quality' and 'service quality' (which are specified in 150 detail-criteria) in order to achieve the top level of four flowers.

Branding

'To become an internationally well known brand in the field of rural tourism' is the top goal as previously outlined. The branding strategy has been found as the most promising way to achieve the development and marketing objectives under the conditions where many small enterprises with limited financial resources (Roberts and Hall, 2001: 205).

In the past for the provincial farm holiday brochure it was deemed sufficient to integrate the logotype somewhere on the cover page. Now the whole outline of the brochure and the most important pieces of information have been designed according to common branding guidelines for farm holidays in Austria. The effect is that the consumer can immediately detect that the individual brochures are part of one product family.

Marketing

The definition of key markets, key target groups and key products has helped to concentrate limited resources. Additionally, the clear decision that niche products cannot be the responsibility of the Federal Association has helped to clarify the competence of each organisational level. Thus lengthy discussions on individual issues can be avoided by referring to the agreed strategy.

In two meetings the joint summer and winter marketing activities are arranged. By the pooling of limited financial resources, major marketing activities such as television promotions, poster campaigns in major Austrian cities, and even the naming of a train (in Austria it is possible to take the patronage of a train) become affordable. All these activities have resulted in a strong increase in personal enquiries to the office of the Federal association (see Fig. 10.1).

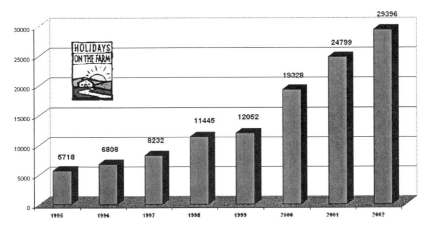

Figure 10.1   Increase in personal enquiries for farm holidays brochures to the Federal Association, 1995-2002

Organisation/Service Centres

In order to concentrate the available human resources for major parts of the Organisation's work, service centres have been formed. This means that from three to five people from the Federal and Provincial Associations work on a special task and develop this area for the whole organisation. The results are finally presented to the decision-making body and made available (or compulsory) for the whole organisation. By March 2003 the following service centres had been created:

Brand service Branding is one of the key tasks as developing a strong brand is a primary goal. The service-centre works out the basics for realising this strategy on all levels. The basic rules and guidelines have to be worked out for all available media (from branded stationery and farm holiday brochures to the appearance of the farms on the internet) and communicated to the members of the organisation as well as to each member landlord.

E-service This group deals with further developing internet presentation guidelines, the farm holidays intranet and other IT applications. In this group the IT 'know how' of the organisation is concentrated, and the individual associations and members benefit from considerable synergies. As far as the effects at the farms are concerned, the internet has become the most important information channel, but increasingly also the sales channel for farm holidays in Austria. Therefore this service centre deals with one of the key strategic issues.

Training and education service The new strategy and challenges for all people involved in the field of farm holidays result in a high need for information, training and education. The contents range from a basic certificate for beginners to specific training for special internet applications. This group works out and organises the courses in close co-operation with the provincial chambers of agriculture (which organise training programmes for all farmers) and the Federal Ministry of Agriculture. Here the farm holidays organisation only determines the necessary contents; the responsibility for training and education lies with the partner organisations.

Quality service Within the scope of this service centre fall tasks like developing the classification system as outlined above, and also ensuring the quality of the work of the organisation, for example how enquiries and complaints are to be dealt with, and how guest surveys or membership surveys will be handled.

Booking service One of the challenges of the closer co-operation is to create a reservation centre for the whole country. To date, a professional reservation system only exists in the province of the Tyrol. However, experience from trade fairs shows that there is considerable interest and demand for immediate bookings (instead of just collecting information or ordering a brochure) as well as a good chance for selling the product actively to the trade.

The service centres described above have resulted in a considerable increase of flexibility and efficiency, but also of the professional quality of the Organisation's work.

Education/Training

The contents of a strategy are by nature sometimes abstract and technical. The challenge is to break the strategic guidelines down into practical steps and activities that can be taken by the individual member or group of farmers. A lot of this 'translation and transformation work' is done in education and training courses.

Lobbying

As an organisation with many operational and financial links to other private or public bodies, effective lobbying for farm holidays – mainly in agriculture and tourism – is important. In the past this was done as and when required. Within the new strategy, the organisation has worked out a systematic lobbying process for the various target groups.

Communication

As not all part-organisations are involved in working out the various actions and products any more, communication has become a crucial issue to ensure acceptance, and consequently the success, of the whole strategy. This concerns the managers and staff members of each part-organisation and, very importantly, the elected representatives of the members in each region who have the task of translating often complex issues to a wider membership, who may not be as deeply involved in tourism as is, for instance, an owner of a hotel.

The main communication instruments are meetings, personal letters, newsletters, the membership magazine, and the internet.

Experiences with the Process

The following points summarise the key experiences arising from the process of working out a new strategy:

- at the outset it had to be communicated that in the interests of future success, sticking to old – if proven – procedures was not an option, and that after new processes were underway, there was no going back;
- the higher involvement and responsibility (e.g. for complete projects) of more members of the organisation (mainly managers of provincial associations) proved to be a valuable and successful change with respect to future development;
- as a result, increased flexibility was achieved which reduced the 'bottleneck' syndrome caused by the unnecessary duplication of common tasks and a very small Federal Association;
- particularly at the provincial level, the more involved persons ('in group') find it much easier to communicate and implement the strategy;
- the involved external partners were interested and contributed actively towards the new strategy;
- it was important to find and use a common language. When farmers, administrators, tourism experts and consultants work together, frequently the same word has different meanings for different groups and in extreme cases no meaning to certain persons;
- the communication of process and results is one of the key elements for the success of the whole work; and
- the systematic work with change during the process, and discussions of alternative solutions, reduced the fear of change in general.

The following difficulties were identified:

- the process of strategy development and formulation is by nature somehow 'academic';
- the landlords and their representatives partly like to pursue a too 'practical approach', strategy work and work for the organisation also requires a theoretical approach;

- the 1989 strategy was simple and easy to communicate, and as many elements of it were new, they were largely theoretical. For the new strategy, the challenge was to further develop and improve an existing successful organisation both theoretically and practically;
- organisational development towards one coherent organisation reduces autonomy for the Provincial Associations;
- realisation of unpopular changes, for example higher contributions to the organisation, were not welcomed;
- the Provincial Associations were in different stages of development (organisational background, financing, objectives). This is a serious challenge for the further development of the whole organisation;
- stronger emphasis on e-services creates 'two classes' of members and the challenge of how to serve members without internet access;
- particularly in the first year, the changed procedures proved rather difficult as the associations had to adhere to new agreements, for example changing their proven way of producing their own brand identities and adopting brochures and systems without a track record; and
- the chosen approach of developing the organisation and at the same time working out a new strategy resulted in a heavy workload especially when projects got underway.

Experiences with the Results

For the motivation of all those involved it was important to have a clear end-date for the two-year development process with decisions and the presentation of results on the occasion of the Federal Association's 10-year anniversary.

During the new strategy development process and its first phase of implementation, it was very helpful that the operational work showed such positive results – a clear increase in enquiries with the organisation and bookings on member farms. This helped to maintain motivation during a time of intense continuing work on the strategy.

When a number of activities are decided upon – and they all need a certain amount of financial and human resources – there is a danger that they are started too ambitiously, there are too many projects running simultaneously, that the 'stress factor' continues, and a feeling occurs that with the new strategy additional work has been created rather than improving the situation.

It is relatively simple to decide on attractive goals ('we will create a strong international brand in the field of rural tourism'). However when it comes to taking the necessary measures to implement such a strategy, some would rather take the 'easy option'.

If there is general agreement in the new strategy concerning a specific strength of one association or a preferred activity of one person, some people find it difficult to give up their own way and will find 'objective' arguments as to why it is ineffective (or impossible) to comply. In such cases it is recommended that the person or organisation concerned becomes involved with the work of the service centre which deals with the task in question.

Experience showed that members were not really interested in theoretical issues, administrative solutions within the organisation, and back-office tools. Presentations and discussions have to be focused on facts and measures that clearly and directly have an effect on them.

Care must be taken that all associations contribute – according to their size and resources – to the co-operative work in the service centres. Experience has shown that not all parts of the strategy can be implemented in the way intended and elaborated in the strategy. It took courage to rethink a few approaches in order to make the work flow again.

If the development of the whole strategy and its specific actions are not decided and communicated incrementally and frequently there is the danger that people feel they are being directed from 'above'/from 'the centre'. Consequently, things are blamed 'on the centre', ignoring the fact that measures being taken are the result of joint effort, and that members of the Federal Association merely have a co-ordinating role.

The content of the written strategy is often formulated in a specific marketing language with technical terms. This is often necessary to be precise and understood in the marketing world. However, in the implementation phase there is a strong need for 'translating' the strategy into the everyday language of the members, to find practical examples and images to explain certain ideas, and to break down the strategy into clear actions and activities at each level.

It was very helpful that these considerable changes could be made with an extremely experienced and stable team on the board. An existing trust in the leading representatives of the landlords, and in their competence and stability, supported this period of change.

It is necessary to communicate clearly to all organisations involved the improvements that have been made – marketing results (for example in terms of increased enquiries); cost reductions (for example through the production of common stationery); more effective working practices (for example the production of common training materials); and examples of reduced workloads (through the use of common membership bases for example).

In general the new and attractively presented strategy gives all persons involved in farm holidays the feeling of being up-to-date again, of having an adequate organisation and modern marketing tools available in times of rapid change. The new strategy has been formulated from the practical experience of the past and helps to maximise limited resources.

Presentation of the Results

The results were presented in several ways:

During the process:
- in 'provincial conferences' as outlined above; and
- in the form of a 'living protocol' on the internet where everybody could make written remarks to the process or the contents in work.

At the end of the process:
- presentation at the 10-year-anniversary of the Federal Association;
- with a presentation-meeting in each province, mainly for decision-makers and 'multipliers';
- with a printed brochure 'Holidays on the farm – the way to a successful brand';
- publication of 'Strategy 2001-2010' which included a summary of analysis, prospects, visionary goals, strategy and motivation for the implementation;
- the provincial associations also received CD-ROMs for their own use; and
- public relations activities in the organisation's membership magazine, in trade media in agriculture and tourism.

The overall objectives were to emphasise the unique strength of a whole sector working with one coherent strategy and the insight of the farm holidays sector in focusing its work on current and future market developments rather than treading well worn paths and perpetuating old working practices.

Conclusions

The strategy development process was an attempt to secure the results of ten years of experiences and successful work for farm holidays in Austria and at the same time to prepare for the future. It was necessary – with the involvement of a number of stakeholders groups – to find ways to become more flexible and to organise resources to meet such future challenges.

The results clearly show that the organisation has successfully completed the pioneer phase and is now on its way to change the familiar procedures to a more professional co-operation as one coherent body rather than nine separate (eight Provincial and one Federal Association) organisations.

This change will be an ongoing process as is the task of developing the strategy for farm holidays in Austria. This will be one of the major challenges for everyone involved in this very interesting sector at the intersection of agriculture and tourism in Austria.

References

Deutsche Reiseanalyse (2003), *German Travel Analysis*, NIT-Institute, Berlin.

Ilbery, B. and Bowler, I. (1998), 'From Agricultural Productivism to Post-productivism, in B. Ilbery (ed.) *The Geography of Rural Change*, Addison Wesley Longman, Harlow, pp. 57-84.

Roberts, L. and Hall, D. (2001), *Rural Tourism and Recreation: Principles to Practice*, CAB International, Wallingford.

Chapter 11

Networking and Partnership Building for Rural Tourism Development

Alenka Verbole

Introduction

Rural tourism development is more than just a planned process. Using an actor-oriented approach (Long, 1989, 1992, 1997; Long and van der Ploeg, 1989; Verbole, 1999; Villarreal, 1992) it can be seen as a dynamic, on-going socially constructed and negotiated process that involves many social actors[1] (individuals, groups and institutions) who continuously reshape and transform it to fit it to their perceptions, needs, values and agendas. Perceiving and responding differently to changing circumstance brought upon them by the development of rural tourism, social actors align themselves with various normative and social interests (Verbole, 1999, 2000; Long, 1989). This means actors will form alliances with different local and external actors to pursue their own social projects sometimes bringing pressure to bear in reshaping and reconstructing the rural tourism development process.

Verbole (1999, 2000) suggests that, in order to be able to understand the rural development process and how it gets transformed, social actors need to be identified, their organising practices (i.e. social networks, family clans, cliques, factions and even more formally constituted groups such as local councils and so on) and processes need to be investigated and there is a need to give attention to the different social arrangements and discursive/normative commitments that emerge from these interactions.

Further, rural tourism is not developing in a vacuum, but it is embedded in a given social, political and historical context. In Slovenia, this implies taking account of the new political and policy reality, although the old hierarchies and structures cannot be ignored, as attitudes do not change as quickly as events (Verbole, 2000).

The research conducted in Pišece, a small rural community in south-eastern Slovenia[2] (Verbole, 1999, 2000), shows that the networks of 'friends of friends' (Boissevain, 1974), namely the local organisational practices (i.e. cliques, family clans and voluntary associations, societies and clubs), play an important role in the rural tourism development process.

By identifying and analysing various networks and other organisational practices it is possible to gain insights into who has the access to decision-making for rural development and who is deciding the outcome of the issues that concern the community – such as the development of rural tourism (Verbole, 1999). The observations presented below were drawn based on an ethnographic exploration[3] of social realities[4] of the rural development process in the given locality. A number of contrasting social settings, such as pubs, community centre, and administrative bodies were chosen to inquire into how social configurations, patterns of social order and organisations are structured (Verbole, 1999: 64-74).

Social Network Linkages and Formation: Actors Organising Social Practices

Networking is understood here in terms of social interactions among various actors. Leeuwis (1991: 113) defined actors' networks as '... flexible and changing sets of social relations between individual and institutional actors that involve material, social and symbolic exchange'. Networks extend through time and space, and so particular interactions can be understood again in the context of a 'network' or chain of previous and future interactions and in different spatial locations'. These networks can be consciously formed, or they can simply exist as informal agreements between people based on multi-purpose social relations (Box, 1990).

Social relations are seen as being of variable length and often changing (dynamic) chains through which not only the resources, but also the power to influence rural tourism development are generated and exchanged.

The concept of power as understood here goes beyond the 'power over others'. It includes the 'use of power to act' and the 'use of power to achieve chosen ends'. Bernstein (1978) argues that power not only relates to the ability to influence others, but also to the ways and strategies various actors use to negotiate the most favourable terms for development (in this case, of rural tourism). Secondly, following Clegg (1989) power should be seen as an outcome of negotiation and not as something one can own.

Schermerhorn (1961) believed that power is conferred by access to different resources, by the identification and defence of particular interests (an interest may be defined as a pattern of demands and expectations), and by the domination of the means of action in the struggle for access to resources and control.

Interpretations from Schermerhorn (1961), Bourdieu (1984), Long (1992) and Clegg (1989) reveal that several types of resources can be used either to advance power or strengthen a power position. These are:

- economic resources (i.e. different types of economic capital, financial means, land, labour, wealth);
- political resources (i.e. political status, formal position in society, legitimacy, legislation, laws); ·
- social resources (i.e. social status and position in society);
- symbolic resources (i.e., ideology, beliefs, value systems, religion, education, specialised knowledge and propaganda); and

- collective resources (power through collective agency, Giddens in Long, 1992).

In the context of the above, Pišece's influential actors and their area of social network linkages and formation (i.e. different structures of influence and power) were analysed following contact with the local pubs (Verbole, 1999: 123-127) using theoretical concepts such as domains and arenas. The term domains refers to areas of social life where encounters between different lifeworlds[5] (Long, 1997) take place, while arenas are defined as social encounters or series of situations in which contests over resources, values and representations take place. The domain becomes an arena when one can observe the struggles (Long, 1997).

The analysis of the village's domains (i.e. domains of politics, family, voluntary associations, societies and clubs and church) provides insights into the various aspects of life-worlds of the villagers of Pišece and the arenas in which they strive for or wield influence (Verbole, 1999). By looking at the domains, the different organisational forms embedded in the local community's daily life, actors' activities in various local associations, societies and clubs, their economic strategies and different social networks are described in order to help visualise locals' access to the social, economic and political space of Pišece.

Pubs - Social and Recreational Locales[6] in Pišece

In Pišece, the pubs (*'Pri Anici'*, *'Viator'* and *'Pri Jašku'*), and an inn (the *'Marelica'* Inn) represented important social meeting places were information is exchanged and opinions are formed. Also, Boissevain (1974) suggested that shops and pubs often develop into regular meeting places for cliques of friends. A view of the inn and pubs' clientele provides an overview of the social organisational practices in the community. Verbole (1999) observed that due to the particular social dynamics of Pišece, the access to these places has been limited or constrained. Some villagers visited a specific pub only while others frequented several according to their specific social group's orientation (Figure 11.1).

The groups in Pišece's pubs had been formed according to the roles that different local actors had played in a dispute more than 15 years old concerning phone installations. The 'privileged' villagers who received their phones first were 'labelled' by those who did not as the *'phone group'*, and the ones that had to wait for their installation called themselves the *'no-phone'* group. Today, when many of the households in Pišece have a phone, including those named the *'no-phone'* households, locals still often use the phone as a frame of reference to distinguish themselves from each other, and they divide accordingly into these groups when visiting local pubs.

The dispute illustrates how past events can be a significant player in the present and that allegiances made or consolidated during particular events can affect present alliances and decisions (Verbole, 1999: 128).

It was also observed that the history of Pišece has influenced present day alliances and allegiances. A social division from another dimension caused by both the philosophical and political allegiances formed between the '*reds*' and the '*blacks*' has had much to do with historical events and with people's religious orientation. (Verbole, 2000).

In Pišece, the '*blacks*' was a label used for people who practised the Christian religion, and favoured the political influence of the Catholic hierarchy or clergy.

This latter term is often used in Slovenia as a synonym for 'conservative', 'traditional' or 'right wing', concepts that Adam (1994) links to what he calls the rural – Christian milieu. The '*reds*', was a broad label used for people believed to be either communist or socialist, as well as for those considered liberal, progressive, or people who are classified in ideological terms as left wing.

All labels (Wood, 1985) are related to social divisions that cut across Pišece's local community and they play prominent roles in locals' lives and consequently in Pišece's attempts to develop rural tourism. Wood (1985: 5) suggests that a focus on labelling can '… reveal processes of control, regulation and management which are largely unrecognised even by the actors themselves'. In Pišece, labels were used to draw symbolic boundaries between locals.

The divisions between the '*blacks*' and the '*reds*', and the '*phone*' and '*no-phone*' groups influenced practically every domain of Pišece's life and even determined which pubs locals frequent, as illustrated earlier.

These social divisions and cleavages proved through time to be crucial factors in the decision-making process and lines of communication in Pišece. The lines of communication – which were defined in Pišece in terms of its history, and 'powers' they might represent – regulate access to different locales, and influence the dynamics within the various domains. Excluding others and sticking to one's 'own' social networks while socialising and discussing serious matters was 'blocking' development.

The local social groups such as family clans, networks, and cliques were found to be very important in obtaining and controlling access to the decision-making process for development. It was through the family clans, networks and cliques that various local actors become involved in strategies to promote, control, reshape and make the most of internal and external interventions. Family clans were built explicitly on kinship ties, while the local networks and cliques were built using other resources, such as religious orientation, political affiliation, value systems, links to actors in positions of power in the local community and to external actors. These factors, importantly, influenced the exclusion and inclusion of actors from certain networks and cliques, thus the rural tourism development process was dominated by the struggles between various groups and ended in a stalemate position at a given time in the process of development (Verbole, 1999).

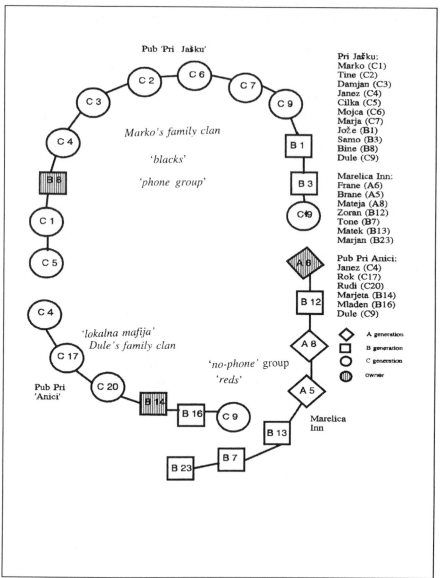

Figure 11.1   Pubs' networks in Pišece

Pišece's Domains – The Basis of Social Life and Influence in Decision-making in the Local Community

The domains are central to understanding how social ordering works (in Pišece) and to analysing how social and symbolic boundaries are created, defended and changed in the local community.

Together with the notion of arena (and how they are bounded), domains provide us with an analytical handle on the kinds of constraints and enabling elements that shape actors' choices and 'room for manoeuvre'.[7] The arenas are either spaces within which disputes associated with different practices of different domains take place (i.e. Pišece's pubs) or they are spaces within a single domain where attempts are made to resolve discrepancies in value interpretations and incompatibilities between actors' interest as will be illustrated below.

What does this mean with regard to rural tourism development, and how important are Pišece's domains with regard to pursuing possible alternatives for rural development? In exploring these issues, the reader is introduced to the domains of family clans, cliques and networks, the domain of voluntary associations, societies and clubs, the Church domain and the political domain.

The Domain of the Family Clans, Cliques and Networks

Organising social practices in this domain involved kin relations, networks of friends, and 'friends of friends', but also included categorisation according to clique alliances (i.e. '*reds*' and '*blacks*', '*phone*' and '*no-phone*' groups), family history and even personal history, as suggested by Villarreal (1994). It became evident during the fieldwork that Pišece was a community dominated by several powerful family clans, each with its extended local and external networks, specific status and access to resources.

Five family clans were identified as the centres of the struggle for power within the process of the development of rural tourism in Pišece (Figure 11.2). In most cases, these clans, based on nuclear family networks, extend to those of their extended family (i.e. grandparents, cousins and so on). Some of the members of the five powerful family clans have, in order to strengthen their position and wield more power, formed a very close knit informal group, called the CRPOV Initiative Board (CIB),[8] that had initiated the change in Pišece. This clique had tried throughout the process to 'monopolise' the development of rural tourism by excluding as many locals – and their family clans and networks – as possible. It seemed that the main criteria for inclusion or exclusion in the clique had to do with whether the 'others' had any links to the five family clans' webs, and whether their networks were extended kin-based or friendship relations, or whether they had resources or power that made them relevant to the clique's aspirations and interests.

While the family clans in Pišece are by definition social groups based explicitly on kinship ties, the cliques and factions are by contrast, informal groupings based on political or ideological orientations and/or common interests. In comparison to the cliques, factions are defined as coalitions or groups of persons (followers) recruited personally according to diverse principles (i.e. religious or political orientation) by or on behalf of persons with whom they were formerly united over honour (respect, influence) and/or over resources. Further, contrary to the cliques, factions are likely to have centrality of focus, thus a leader (Boissevain, 1974).

In Pišece several cliques could be observed and there were clear lines or boundaries drawn between them. Belonging to a certain clique, and thus having access to the clique's networks and, consequently, the relevant resources and information, seemed to be an important factor in determining whether and to what extent one could participate in Pišece's development.

The Five Family Clans and their Extended Networks

The study presented here looked into how these social organisations and practices emerged, how they were used, how new ones were built, i.e. to penetrate various local domains and how they tried to exclude other actors.

It has been confirmed that one's political affiliation in Pišece influenced who one associated with personally (i.e. marriage, friendship, loyalty, etc.) – meaning people grouped together according to their political affiliation, religious orientation, kinship and so on (Verbole, 1999).

To illustrate, the younger generation of Marko's[9] clan (Figure 11.3), one of the five most powerful clans, has been actively involved in Pišece's attempts to create better living and working conditions by trying to enrol in the National Programme for Rural Development known also as CRPOV (Verbole, 1998, 2000). Four of the Marko cousins joined the CIB, while the rest of his kin supported its initiative through positions that they occupied in various formal (i.e. local government) and informal (i.e. voluntary organisations) structures. Thus, Marko's family clan increased its power to influence decision-making and extended its 'room for manoeuvre' by guaranteeing its presence on relevant local and meso level formal decision-making bodies. Other family clans used either their symbolic resources (specialised knowledge) or political resources (such as a family member in the local council), as well as links to actors that had access to external actors' networks (such as university networks or development agency networks), to establish themselves as a key-actors in the development of Pišece. Actors used family clan networks to increase power within other domains, such as those of politics and of voluntary organisations, societies and clubs.

It was noted that clans and cliques would not easily accept people from another family clan or clique. However, the narrow family clan orientation was 'exceeded' (the actors saw and acted beyond the family clan or clique boundaries) when the young people of Pišece formed the CIB, including three members of the older generation (Verbole, 2000). This 'partnership' was based on locals' common interest in creating better conditions for their future rather than focussing on their family clan orientation or any other division or cleavage that cross-cut Pišece. As noted, the voluntary associations, societies and clubs played an important role in this development. The young people knew each other and socialised together through activities organised by various voluntary associations, societies and clubs.

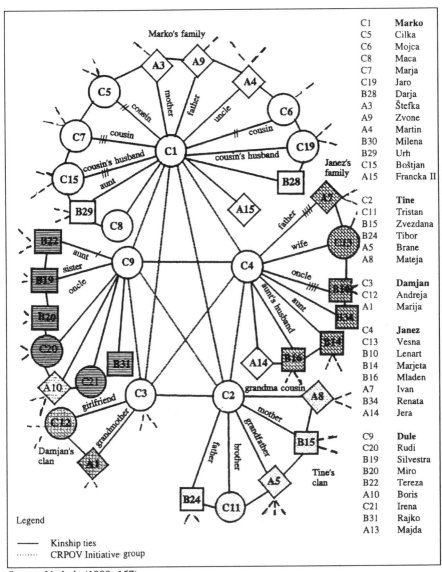

C1	**Marko**
C5	Cilka
C6	Mojca
C8	Maca
C7	Marja
C19	Jaro
B28	Darja
A3	Štefka
A9	Zvone
A4	Martin
B30	Milena
B29	Urh
C15	Boštjan
A15	Francka II
C2	**Tine**
C11	Tristan
B15	Zvezdana
B24	Tibor
A5	Brane
A8	Mateja
C3	**Damjan**
C12	Andreja
A1	Marija
C4	**Janez**
C13	Vesna
B10	Lenart
B14	Marjeta
B16	Mladen
A7	Ivan
B34	Renata
A14	Jera
C9	**Dule**
C20	Rudi
B19	Silvestra
B20	Miro
B22	Tereza
A10	Boris
C21	Irena
B31	Rajko
A13	Majda

Legend

—— Kinship ties
········ CRPOV Initiative group

Source: Verbole (1999: 157)

Figure 11.2   The five family clans and the CIB partnership

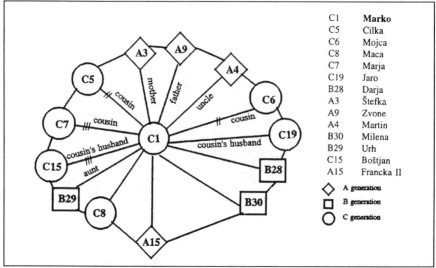

Source: Verbole (1999: 147)

Figure 11.3   Marko's family clan

The Domain of Associations, Societies and Clubs

An examination of Pišece's associations, societies and clubs shows how social ties cut across the various domains in this small community (Figure 11.4). Different associations, societies and clubs had their own expectations for rural tourism development and their own resources to influence decisions related to it. Their expectations and interests had much to do with the expectations and interests of individual members or groups of members, who pursued networks in organisations of particular interests to their clique or family clan (some local actors found it important to enrol in the Tourist Society[10]).

Examining villagers' links with different associations, societies and clubs allowed an investigation into how the informal structure of the local community relates to different actor activities within these organisations. Participation in a given voluntary association, society or club's activities opened access to other networks (and domains) in Pišece, since some actors were affiliated with several organisations giving them the opportunity to extend their social networks.

One of the most popular among the fourteen organisations with regards to its role in the development of the community and the most active was the local Tourist Society, which had been operating in Pišece for more than thirty-five years and everybody was welcomed as a member. Through the years, the Tourist Society had motivated locals to beautify Pišece (houses, farmyards and so on) and to enrol in several national competitions (i.e. for the most beautiful small settlement awarded by the Slovenian Tourist Association). The Society also organised ethnographic exhibitions, culinary and wine-growing related festivities and competitions, flower and hunting exhibitions and competitions.

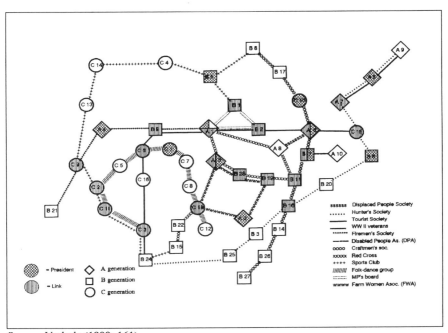

Source: Verbole (1999: 161)

Figure 11.4   Participation of local people in Pišece's associations, societies and clubs

The Hunters' Society was considered by many villagers to be a prestigious organisation providing fun and relaxation for politicians, businessmen, rich Italian and Austrian hunters and the like. The Society was, thus, involved in rural tourism as a matter of course. Compared to the Tourist Society, the Hunters' Society was relatively closed and access to the Society and its own networks was limited.

The various associations, societies and clubs were dominated by the social divisions mentioned earlier. Some local voluntary associations, societies and clubs were more balanced in terms of social divisions between 'reds' and 'blacks' compared to others. The Sports Club, for example, was dominated by the 'reds,' while the folk-dance group and other cultural associations, societies and clubs were dominated by 'black' membership. The 'blacks' had a more conservative grounding in 'traditional' aspects of rural Slovenian life and tended to be less open to a progressive change towards modernity. 'Reds' were more progressive. This implies that voluntary associations, societies and clubs could be used to focus on achieving goals of a particular social group rather than the interests of the voluntary associations, societies and clubs themselves or the wider local community.

On the other hand, it was observed that participation in voluntary associations, societies and clubs allowed locals to enter into new social networks that cross-cut the entire community, thus enabling actors to extend their networks into other networks to gain information and support for their projects, both of a general nature and those related to rural tourism development.

Therefore, some actors were or tried to get involved in as many of the 'important' associations, societies and clubs as possible. 'Important' was perceived as those associations, societies and clubs that were either useful in terms of their potential to influence decision-making in the Local Community[11] (i.e. due to their position in the Local Community) or in reference to gaining access to those 'powerful' people who happened to be members.

Secondly, some actors have, in order to increase their room for manoeuvre in the local power structure, strived to occupy as many key positions as possible (i.e. President, Secretary) in the various voluntary associations, societies and clubs (Verbole, 1999).

The Church Domain

The activities within Pišece's Church domain mainly focused on religious practise, however it also involved organising and participating in local social activities such as the 1994 workshop[12] organisation, the organisation of the Centenary Anniversary of Maks Pleteršnik's Dictionary and the farm youth games.

Practising religion was primarily a '*blacks*' activity, while '*reds*' had hardly anything to do with religion. On the other hand, some social activities organised by the local priest, thus by the Catholic Church, have also been joined by locals with no religious affiliation, and who were labelled '*red*'. For example, the owner of the local tailoring business donated T-shirts to '... the priest's project'.[13] Further, a strong and active participation by the local priest and the Catholic Church was observed in several local (horizontal) as well as external (vertical) networks. The priest participated in the group initiation process of some of the newly emerging informal groups and partnerships in Pišece, i.e. the CIB, as well as playing the role of intermediary – a social actor that bridges the gaps between various networks. The Church's involvement with this as well as some other initiatives raised many eyebrows in Pišece, since many believed that a 'man of the cloth' should not be so active in the broader spectrum of community activities.

In particular, in past times, the Church was unlikely to get involved with 'secular or worldly affairs' such as community development. In Slovenia, the Church was and still is separate from the State. The increased participation of the Church in rural development was made possible first by the shifts towards a multi-party democratic system of government, and also by a liberalisation of culture, suggesting that, in the future, new players in community development processes would have to be reckoned with (Verbole, 1999, 2000).

The Political Domain of Pišece

Interactions within this domain focused on the functioning of the Local Community, *krajevna skupnost* – the nominal political power – and its representatives. The community power structure is defined as a pattern of relationships among the individuals and groups in the community which enables them to influence community decisions on a given issue (Tait, 1976). In the case of Pišece, this implied the need to look closer at the social interactions and relations within the domain of politics and especially at the relationships between formal (the local administrative authorities) and informal structures and their 'power' to influence Pišece's community decisions (and actions) on the issue of most importance to the locals – the issue of 'saving Pišece from dying'.

The nominal political power vested in certain officials in the location in question (often referred to as positional power: Tait, 1976) was counterpoised by the 'actual' or manifest power structure(s) as seen in the confidence and perspectives of the 'administered' – the villagers of the Local Community of Pišece (Verbole, 1999). The struggles that took place within the domain (the informal structures trying to overtake formal positions) as time passed by were followed up.[14]

A distinction between two notions needs to be made here, namely that of 'authority' and that of 'power'. Mendell (in Barnes, 1986) linked the notion of 'authority' to legitimacy and legally established authority representing macro actors (i.e. the State) and imposing regulations and development directions. The notion applies to the actors that occupy social positions that historically carry legitimacy, argument and a 'structure of domination' (relates to the issue of leaders and followers, hierarchy). The field data show that these 'structures of domination' were well embedded in Pišece.

The President and local formal authorities gave little impetus to the creation of opportunities either for the development of the Pišece Local Community, or for the development of rural tourism. Most decisions regarding the development of Pišece were not made by the local government, but rather outside the Local Community President's office (Verbole, 1999) confirming that power is not a fixed structural property, but is negotiated by social actors and can be extremely fluid (Arce *et al.*, 1994; Villarreal, 1994; Verbole, 2000).

Various reasons were identified for the shift in power encountered, ranging from the reform of local government,[15] and the transition to a multi-party democratic system of government (which brought to the surface old divisions between the '*blacks*' and '*reds*' and poured more oil on the fire since '*blacks*' suddenly had lots of potential to access the political domain, thus to 'power') to the lack of access to relevant social networks – which in Pišece determined whether one could achieve things or not. The President, for example, was often considered an 'outsider' as he was not born locally and he also lacked family networks support.

Overall, the study in Pišece showed that most power lay with the people who had access to both local and external networks. Such actors (e.g. local priest) were looked upon for guidance and were used as a primary frame of reference for other villagers, providing them with resources, thus power to influence the direction of rural tourism development. These key-actors were functioning as developmental power brokers[16] – conduits through which information on rural tourism was obtained, filtered and spread, thus controlled and manipulated (Verbole, 1999: 179-180). These power brokers were motivators or de-motivators of the different initiatives by virtue of their power status and/or position.

Likewise, some local actors functioned as gate-keepers.[17] They not only opened the gates to external networks for other locals and resources, but also opened gates for external forces to penetrate the local community and establish themselves within it (i.e. research institutions and private development agencies). Many actors used rural tourism development to a greater extent for their personal rather than collective objectives (i.e. to advance their position of power) and objectives of a more political nature (i.e. Municipality and State).

Both brokers as well as gate-keepers often determine whether other social actors would be included or excluded from negotiating the development of tourism by blocking access to other networks and its resources (Verbole, 1999). As pointed out by Wolf (in Long, 1972: 11), the study of actors 'that stand guard over critical junctures or synapses of relationships which connect the local systems or *networks* to the larger whole' also allows us to comprehend more fully the mechanisms and social processes through which local rural areas are socio-economically and politically integrated within the wider context.

Discussion and Conclusions

The case study from Slovenia shows that access to the rural tourism development process in a given locality is not equally guaranteed to everyone. The local dynamics of social interaction and power relations play an important role in regulating participation in the process.

Actors' social networks influenced social practices. Especially, family clan networks were found to be influential in determining whether one will have access (and up to what level) to certain locales and domains where issues are discussed, commitments shouldered, loyalties and opinions shared or not and, consequently, where opportunities to wield influence are developed.

In the study area, the contrast of social constructions (perceptions, value systems, kinship, religious and political affiliation) in opposing local social groups was used as a main 'sieve' for exclusion and inclusion into the networks and other social organisational forms.

As the process of social change continued to be negotiated and transformed, different domains have been very important in actors receiving and exchanging information as well as pursuing possible alternatives in the development of rural tourism, as they were less burdened by social segregation (in particular the domain of the voluntary associations, societies and clubs).

Overall, it can be stressed that social dynamics can potentially have an impact on present as well as future developments and that social divisions and cleavages are crucial factors in decision-making and lines of communication, therefore should be taken into consideration by policy planners and developers.

Notes

1   The term social actor is restricted to those social entities that can reach decisions and act accordingly (Long, 1989). Thus, the concept of social actor should not be used to cover collectives, agglomerates or social categories that have no discernible way of formulating or carrying out decisions (Long, 1989).

2   The site of the case study presented was in Pišece Local Community, a small community of 1200 inhabitants in the Southeast of Slovenia that requested assistance to develop their local community. For more insights see Rus, 1995; Verbole, 1997, 1999 and 2000.

3   Ethnographic data were gathered according to the grounded theory approach (Strauss and Corbin, 1990) during three years of qualitative field research (1994-1997) using informal and semi-structured interviews (i.e. with local community members in both formal and informal positions of authority), life histories (i.e. of key family clan members) (Long and Roberts, 1984), extended case studies (Mitchell, 1983), participatory observation (Bernard, 1988) and a situational analysis of critical events and issues (Long, 1992 and 1997). Theoretical and methodological approaches taken are discussed in more detail in Verbole, 1999 and Verbole and Cotrell, 2002.

4   Different social actors have different perceptions of reality. These multiple perceptions and opinions – multiple realities (Long 1984, 1988) influence the way actors respond to changing circumstances (e.g. by the rural tourism development project/process). Actors' respective realities also influence how they align themselves with normative and social interests involving power, authority, and legitimisation (how they develop strategies and responses to a given situation), which may be just as likely to cause conflict among social groups as they are to contribute to the establishment of common ground.

5   Lifeworld is defined as an actor-specific set of motives, meanings, emotion, daily action, and behaviour (Long, 1989). The lifeworld is defined by the actor who lives it, rather than by any observer (i.e. a developer or a researcher). The notion of lifeworld allows us to understand 'how people move through life', and enables us to learn about the types and contents of important social relations and activities that involve an actor.

6   'Locales' provide for a good deal of fixity underlying institutions, although, there is no clear sense in which they determine such fixity (Giddens, 1984).

7   Long (1989) argues that the notion of 'room for manoeuvre' is central to understanding the ways in which actors attempt to create space for pursuing their own social projects, or for resisting the imposition of other more powerful actors. In Pišece, through the 'room for manoeuvre' that local actors created for themselves, they were able to directly influence some of the activities related to the development of rural tourism at the local level.

8   The CRPOV Initiative Board (CIB) was formed in 1994 by a group of young intellectuals (recent graduates and students) and the three more 'senior' members of the local community, namely the President of the local Tourist Society, the local parish priest and the head of the local primary school (Verbole, 1999, 2000). The CIB's main 'collective' objective was to ensure that Pišece should get rid of its peripheral image, and that young people should have the opportunity to live and work in Pišece.

9   The informant's code is C1. Following Wegner's (1983) suggestion on research implementation, names of informants were changed to protect the confidentiality of sources of data, and the integrity of the interviewees. The coding system for local actors works as

follows: the letters 'A', 'B' and 'C' indicate the actor's age ('A' stands for villagers who are 60 and older, 'B' stands for those who are between 40 and 60, 'C' for those between 20 and 40). The numbers are used to distinguish between various individual actors of the same age group. A similar coding system has been developed for external actors (Verbole, 1999).

10 Tourist Society is a local branch of the Slovenia's Tourist Association (TZS), a non-governmental, non-political organisation. TZS was established in 1905. The goals and tasks of TZS range from stimulating the development of tourism in co-operation with private and public sectors to nature preservation, promotion and education (Verbole, 1999: 243).

11 The term Local Community is used as a synonym for *krajevna skupnost*, the sub-municipal local administrative area that is usually made up of several neighbouring settlements.

12 An International Workshop on Rural Development in Peripheral Areas was organised jointly by the University of Ljubljana and the University of Klagenfurt in 1994 in a well-known resort close to Pišece (Verbole, 1995, 1997). The Workshop allowed a first glimpse of the people and situations in the locality as well as complexities likely to be involved in the many interfaces and discourses of tourism development at the local level (Verbole, 1999).

13 The 'successful' involvement of the local Catholic Church in Pišece had much to do with the priest and his agency; his ability to influence others and enrol them in his projects created room for manoeuvre for himself as well as the Catholic Church and Pišece (see Verbole, 1999).

14 The official power structure – represented by the president, secretary and the council of the Local Community in Pišece – was not only weak, but also 'falling apart' and the growing influence of informal structures such as the role of the CIB in the initiation of the 1994 Workshop trying to wield power could be observed (Verbole, 2000).

15 The reform of local government in 1994 brought with it lots of change at the *krajevna skupnost* and municipal levels. This, among other factors, brought into question the role of the *krajevna skupnost* and its formal body of authority – the Local Council, including that of the President of the Local Community.

16 As power brokers, actors can sanction developments, suggest new ideas, and provide resources or access to resources, both inside and outside the community. Villarreal (1994) suggests that these actors may use their influence directly, to control economic and other resources, or indirectly, to manipulate strategic contacts with people who control such resources, or have access to influential people and relevant political networks, thus to the gate-keepers. Eventually, this control may obstruct specific groupings in participating in the rural tourism development process and limit their opportunity to benefit from development (Verbole, 1999).

17 The notion of 'gate-keeper' refers to actors representing 'linkages' to other networks, thus indirectly to their resources and power (e.g. Damjan as a link to the University). A gate-keeper is in fact, a communication channel between different social groups and/or levels, and does not necessarily occupy key positions in the community's formal as well as informal power hierarchy. The term 'gate' refers to an interface (usually one actor) between the two networks or two 'worlds'. More importantly, the 'gate-keeper' does not manipulate his/her resources. The study from Pišece shows that this concept is dynamic because it is embedded in a wider social, economical and political context (e.g. the new position of '*reds*' and '*blacks*' in Pišece and Slovenia). That implies that if a context changes, so will the actors that occupy the position of gate-keepers. Data from Pišece show that, while in the past it was important to know someone who belonged to the 'old socialist boys network', nowadays it seems important to forge relations with people that belong either to the 'right' political parties or people with friends in the right societal positions (e.g. white collars in ministries). Gate-keepers can become brokers. The term 'gate-keeping' refers to the ability

of actors to accept information and pass it to other actors. As soon as a gate-keeper manipulates (filters or discards) the information, thus restricting other actors' awareness and access to all the information, she/he becomes a (information and/or power) 'broker' (Verbole, 1999).

References

Adam, F. (1994), 'After Four Years of Democracy: Fragility and Stability', in F. Adam and G. Tomc (eds), *'Small Societies in Transition: The Case of Slovenia'*, Slovene Sociological Association and Institute for Social Sciences, Ljubljana, pp. 34-50.

Arce, A., Villarreal, M. and de Vries, P. (1994), 'The Social Construction of Rural Development: Discourses, Practices and Power', in D. Booth (ed.), *Rethinking Social Development: Theory, Research and Practice*, Longman, Harlow, pp. 152-71.

Barnes, B. (1986), 'On Authority and its Relation to Power', in J. Law (ed.), *Power, Action and Belief: A New Sociology of Knowledge?*, Routledge, London, pp. 180-95.

Bernard, H.R. (1988), *Research Methods in Cultural Anthropology*, Sage, London.

Bernstein, R. (1978), *Restructuring of Social and Political Theory*, University of Pennsylvania Press, Philadelphia.

Boissevain, J. (1974), *Friends of Friends: Networks, Manipulators and Coalitions*, Blackwell, Oxford.

Bourdieu, P. (1984), *Distinction: A Social Critique of the Judgement of Taste*, Harvard University Press, Cambridge.

Box, L. (1990), *From Common Ignorance to Shared Knowledge: Knowledge Networks in the Atlantic Zone of Costa Rica*, Wageningen Agricultural University, Sociological Studies, vol. 28, Wageningen.

Clegg, S.R. (1989), *Frameworks of Power*, Sage, London.

Giddens, A. (1984), *The Constitution of Society: Outline of the Theory of Structuration*, Polity Press, Cambridge.

Leeuwis, C. (1991), 'Equivocations on Knowledge System Theory: An Actor-oriented Critique', in D. Kuiper and N. Rölling (eds), *The Edited Proceedings of the European Seminar on Knowledge Management and Information Technology*, Wageningen Agricultural University, Wageningen, pp. 107-16.

Long, N. (1972), *Kinship and Associational Networks Among Transporters in Peru: The Problem of the 'Local' Against the 'Cosmopolitan' Entrepreneur*, Paper presented at the seminar on 'Kinship and Social Networks', Institute of Latin American Studies, London University, London.

Long, N. (1984), *Creating Space for Change: A Perspective on the Sociology of Development*, Wageningen Agricultural University, Wageningen.

Long, N. (1989), *Encounters at the Interface: A Perspective on Social Discontinuities in Rural Development*, Wageningen Agricultural University, Sociological Studies, vol. 27, Wageningen.

Long, N. (1992), 'From Paradigm Lost to Paradigm Regained: The Case for an Actor-oriented Sociology of Development', in N. Long and A. Long (eds), *Battlefields of Knowledge: The Interlocking of Theory and Practice in Social Research and Development*, Routledge, London, pp. 16-43.

Long, N. (1997), 'Agency and Constraint, Perceptions and Practices. A Theoretical Position', in D. de Haan and N. Long (eds), *Images and Realities of Rural Life: Wageningen Perspectives on Rural Transformations*, Van Gorcum, Assen, pp. 1-20.

Long, N. and Roberts, B. (1984), *Miners, Peasants and Entrepreneurs: Regional Development in the Central Highlands of Peru*, Cambridge University Press, Cambridge.

Long, N. and van der Ploeg, J. (1989), 'Demythologising Planned Intervention: An Actor Perspective', *Sociologia Ruralis*, vol. 29, pp. 226-49.

Lukes, S. (1974), *Power: A Radical View*, Macmillan, London.

Mitchell, J.C. (1983), 'Case and Situation Analysis', *Sociological Review*, vol. 31, pp. 187-221.

Rus, A. (1995), 'Socio-geographic Characteristics of Pišece: Problems and Potentials of Peripheral Settlements', in A. Barbič and D. Wastl-Walter (eds), *Sustainable Development of Rural Areas: From Global Problems to Local Solutions*, Klagenfurter Geographische Zeitung 13, Klagenfurt, pp. 117-25.

Schermerhorn, R.A. (1961), *Society and Power*, Western Reserve University, New York.

Strauss, A. and Corbin, J. (1990), *Basics of Qualitative Research: Grounded Theory Procedures and Techniques*, Sage, London.

Tait, J.L. (1976), *Identifying the Community Power Actors: A Guide for Change Actors*, U.S.D.A., Washington.

Verbole, A. (1995), 'Tourism Development in the European Countryside: Costs and Benefits', in A. Barbič and D. Wastl-Walter (eds), *Sustainable Development of Rural Areas: From Global Problems to Local Solutions*, Klagenfurter Geographische Zeitung 13, Klagenfurt, pp. 60-75.

Verbole, A. (1997), 'Rural Tourism and Sustainable Development: A Case Study on Slovenia', in H. de Haan, B. Kasimis and M. Redclift (eds), *Sustainable Rural Development*, Ashgate, Aldershot, pp. 197-215.

Verbole, A. (1998), 'Negotiating Terms of Tourism Development in the Countryside', *Environnement & Societe, Innovations Rurales*, no. 20, pp. 51-59.

Verbole, A. (1999), *Negotiating Rural Tourism Development at the Local Level: A Case Study in Pišece, Slovenia*, Wageningen Agricultural University, unpublished PhD Thesis, Wageningen.

Verbole, A. (2000), 'Actors, Discourses and Interfaces of Rural Tourism Development at the Local Community Level in Slovenia: The Social and Political Dimensions of Sustainable Rural Tourism Development', *Journal of Sustainable Tourism*, vol. 8, pp. 479-90.

Verbole, A. and Cotrell, S. (2002), 'Rural Tourism Development: Case of a Negotiating Process in Slovenia', *WLRA Journal*, vol. 4, pp. 21-29.

Villarreal, M. (1992), 'The Poverty of Practice: Power, Gender and Intervention from an Actor-oriented Perspective', in N. Long and A. Long (eds), *Battlefields of Knowledge: The Interlocking of Theory and Practice in Social Research Development*, Routledge, London, pp. 247-67.

Villarreal, M. (1994), 'Wielding and Yielding: Power, Subordination and Gender Identity in the Context of a Mexican Development Project', Wageningen Agricultural University, unpublished PhD Thesis, Wageningen.

Wegner, G.C. (1983), *The Research Relationship: Practice and Politics in Social Policy Research*, Allen & Unwin, London.

Wood, G. (1985), *Labelling in Development Policy: Essays in Honour of Bernard Schaffer*, Sage, London.

Chapter 12

Integrated Quality Management in Rural Tourism

Ray Youell

Introduction

Few rural tourism professionals would disagree that delivering quality is a key requirement of meeting visitors' needs and, thereby, securing business advantage in an increasingly competitive market place. The concept of integrated quality management (IQM) has only recently been applied to the tourism industry, initially through a series of projects sponsored by the European Commission focusing on urban, coastal and rural tourism. This chapter introduces the concept and principles of IQM as applied to the rural tourism sector and outlines new research into the introduction of the IQM concept and principles in three pilot areas in the county of Ceredigion in West Wales. With the assistance of European Structural Funds, a project officer has been working with the three communities since the beginning of 2002 to introduce the IQM concept and develop strategies to improve quality in all aspects of tourism development, promotion and monitoring. This chapter illustrates the process, opportunities and challenges offered by establishing integrated quality management in rural tourism and concludes that IQM has a pivotal role to play in improving the competitive position of the rural tourism sector while at the same time safeguarding social, cultural and environmental integrity.

The Evolution of Rural Tourism

Many of the world's rural areas are in transition. Prices of land and agricultural produce are falling, many retail and other community services are in decline, and just one of a whole new series of opportunities that rural communities can consider (Long and Lane, 2000).

Lane (1994) chronicles the historical context in the development of rural tourism and examines some of the key issues that combine to make rural tourism distinctive, highlighting the impacts of changes in rural tourism since the 1970s. Within the literature on rural tourism, a range of themes and issues emerges, most notably the impact of rural tourism (social, cultural, economic and environmental impacts), research on different forms of rural tourism (e.g. farm tourism) and the implications for rural areas of a policy of tourism development (Page and Getz, 1997). Farm tourism features strongly in the literature and is often used interchangeably with rural tourism, although it should be remembered that the former is but one 'product' offered within the rural tourism sector. Busby and Rendle (2000) suggest that farm tourism needs to be seen in the wider context of rural tourism given that it forms a key component of both the accommodation supply and many of the attractions available.

Long and Lane (2000) postulate that the development of rural tourism has been through what could be called 'phase one' of its evolution. Across the world, communities and enterprises have begun to practice the art of tourism in the countryside. There has been a steady, if not spectacular, development of the rural tourism product across the globe, with many agencies and individual enterprises involved in policy formulation, product development, marketing, financing, business support, training and enterprise management, with varying degrees of collaboration and integration. Long and Lane (2000) suggest that the next phase of rural tourism development will be a much more complex one, focusing on expansion, differentiation, consolidation and understanding. They list ten implications that are likely to occur in 'phase two' of rural tourism development, including increased competition, more rural tourism policies, partnership building, growth in the market, more effective and sophisticated marketing, product development, training, real estate issues, development around heritage and more sustainable tourism policy. To this list could be added the need for greater integration within the rural tourism sector, both in terms of integrated management practices in enterprises and the appreciation that tourism has a valuable contribution to make to sustainable rural development in its widest sense.

What is clear from a review of the literature is that much has been written about individual aspects of rural tourism in a variety of national and regional settings, but there is little evidence of the need for, and potential benefits of, an *integrated* approach to rural tourism development. It is true that partnerships, representing one model of integration, are receiving growing recognition as a means of encompassing the often disparate components of rural tourism development. This has led to increasing attention being directed to the use of collaborative arrangements or partnerships that bring together a range of interests in order to develop and sometimes also implement tourism policies (Bramwell and Lane, 2000).

The issue of *quality* of the rural tourism sector generally, and the rural tourism product in particular, does not feature strongly in the literature. Up to now the management of quality in destinations in relation to competitiveness has received very little attention (Ritchie and Crouch, 1997). From an integrative perspective, there appears to be considerable room for improvement in developing integrated quality management for tourist destinations and thus increasing competitiveness (Go and Govers, 2000). This chapter suggests that applying integrated quality management (IQM) to rural tourism destinations has the potential to improve competitiveness within the sector while at the same time safeguarding social, cultural and environmental integrity.

The Drive for Quality in Rural Tourism

In many respects, the pursuit of quality in all aspects of the rural tourism product is no different from that of any other sector of the tourism industry. All tourism operators, regardless of the sector in which they operate, are well aware of the competitive nature of the tourism business environment. They also need to be mindful of the importance of determining customers' needs and providing high quality services and products to satisfy their requirements. As Go and Govers (2000) remark, destinations are increasingly reliant on the delivery of quality products and services, and where customer needs and business goals are increasingly inseparable, every enterprise in a destination, not just its public management, must be committed to meet customer needs. Quality has become a major interest of private and public operators in the emerging global market (Go and Govers, 2000).

Much of the research to date on quality in tourism has focused on service quality in the hospitality sector and, in particular, measurement of customer perceptions of quality (see, for example, Augustyn and Ho (1998), Diaz, Vazquez and Ruiz (1998), and O'Neill (2000)). There is little in the literature on *integrating* quality into all aspects of managing tourism enterprises, for example delivering quality in marketing, product innovation, information provision, accommodation, events and outdoor recreation opportunities, to name but a few. Without attention to integrating quality into all aspects of planning, managing and evaluating tourism products, the visitors' experience is unlikely to be optimised.

At policy level in the UK context, the government's commitment to quality in tourism was signalled in its first, fully-fledged tourism strategy, in which it pledged to help the tourism industry turn itself into one that puts quality at the heart of everything that it does (Department for Culture, Media and Sport, 1999). In Wales too, quality is seen as pivotal to developing and sustaining a world-class tourism industry in the twenty-first century, being one of the four key principles on which the Wales Tourist Board's strategy is based, along with sustainability, competitiveness and partnership (Wales Tourist Board, 2000).

Although espoused by many public sector agencies, quality in rural tourism does not feature significantly in the literature. It is, however, a key theme in recent UK strategies on sustainable tourism (English Tourism Council, 2001a) and rural tourism (English Tourism Council, 2001b). In seeking to realise its vision for sustainable tourism, the English Tourism Council places 'delivering a quality tourism product' at the top of its agenda. Quality features strongly in the four aims of the rural tourism strategy, which are:

1   to maintain and increase the availability and quality of employment in rural tourism enterprises;
2   to ensure that a high quality of visitor experience in the countryside is available to everyone;
3   to maintain and enhance the quality of the rural environment; and
4   to spread the benefits of tourism throughout rural communities.

(English Tourism Council, 2001b)

The rural tourism strategy calls for an *integrated* approach to achieving these aims, recognising that they are mutually reinforcing and that many actions can be taken that help to meet two or more, sometimes all four.

At the heart of the debate about quality in rural tourism is the notion of devising standards against which enterprise and individual performance can be measured for the benefit of customers. Consumers should be better informed about distinct rural opportunities and the special developmental and management standards needed to deliver high quality, sustainable product (Page and Getz, 1997). Rural products, especially accommodation and visitor attractions, are frequently included in grading systems that seek to offer customers an objective assessment of quality standards. Attempts have also been made to implement specific rural accommodation grading systems. Notable in this regard was the successful Farm Tourism Award developed in the early 1980s by the Wales Tourist Board. This scheme was a combination of minimum standards for accommodation and services, coupled with a compulsory period of training for operators, covering a variety of topics including customer service, legal aspects of business operation and marketing. However, there are difficulties in devising and implementing uniform quality standards in a sector such as rural tourism, whose very appeal is, in many respects, its lack of uniformity founded on a diverse accommodation stock, disparate events and local/regional specialisms. In short, it is difficult, if not impossible, to compare the quality of service and facilities offered by, say, a farm guesthouse in the Cotswolds with that of a country house hotel in Ireland or a castle in Wales. As Page and Getz (1997) comment, standards should recognise the innate special qualities of rural tourism and the need to preserve the environmental appeal of rural areas. In the UK context, one company that has successfully offered a high quality rural tourism product is Welsh Rarebits, a collection of very different hotels, guesthouses and inns in Wales aimed at the UK and overseas market.

Integrated Quality Management in Rural Tourism

An extension of the principle of total quality management (TQM) so often favoured by the manufacturing sector, integrated quality management is a tool that has the potential to help rural destination managers achieve their twin objectives of increasing local income and employment through tourism, while at the same time ensuring that the environment, culture and quality of life of local people are not damaged by tourism. The adjective 'integrated', meaning to amalgamate, is appropriate in that tourism consists of an assembly of elements, enabling the temporary migration from a 'usual habitat' to one or more destinations for business, recreational or other purposes (Go and Govers, 2000).

The European Commission initiated a series of projects in the late 1990s to develop strategic guidelines based on practical experience to improve integrated quality management in tourism destinations. The working definition of integrated quality management used in the projects was '...an approach to managing a tourism destination which focuses on an ongoing process of improving visitor satisfaction, while seeking to improve the local economy, the environment and the quality of life of the local community' (European Commission, 1999b). The projects investigated coastal, rural and urban tourism through a series of case studies of good practice (European Commission, 1999a, 1999b, 1999c). If we accept the assertion contained within the IQM project reports that quality exists only to the extent that a product or service meets the customer's requirements and expectations (European Commission, 1999b), it stands to reason that the individual elements of a rural tourism strategy based on quality standards must be founded on a thorough understanding of customer needs and expectations. Herein lies the basis of integrated quality management in rural tourism, which has the following two key elements:

1   focusing on visitors – improving the quality of what is provided for them, satisfying their needs and influencing their activities, so that they come back again and/or recommend others to do so;
2   involving local communities and local tourism enterprises in the management of the destination, as participants and as customers of the management process.

It could be argued that there is nothing new in attempting to achieve these twin objectives of improving visitor satisfaction and involving local communities in the tourism development process; what sets IQM apart is the *integrated* nature of the process. While there are examples of success in raising quality in individual elements of UK rural tourism, such as projects concentrating on effective marketing (e.g. FarmStay UK), customer service initiatives (e.g. Welcome Host) or information provision (e.g. the development of rural tourism websites), IQM seeks to improve quality across a range of elements at the same time in an integrated manner within the destination; these include:

• marketing and communication;

- visitor welcome, orientation and information;
- accommodation;
- local produce and gastronomy;
- attractions and events;
- countryside recreation;
- environment and infrastructure.

Figure 12.1 demonstrates the process of integrated quality management in rural tourist destinations, illustrating the three key stakeholders in the process (community, visitors and tourism enterprises) and the intended beneficial outcomes of the process (satisfied customers, improved enterprise performance, increased income and employment, community benefit without conflict).

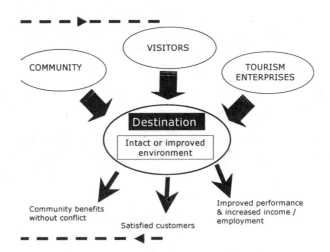

Source: European Commission (1999a)

Figure 12.1   The integrated quality management (IQM) process

Integrated quality management in rural tourism is designed to be an all-embracing concept, addressing both the internal, operational and management issues specific to individual enterprises, while at the same time tackling the wider aspects of social, cultural and environmental impacts of tourism.

Table 12.1 lists the core principles of IQM, which, when taken together, offer a blueprint for the sustainable development of rural tourism based on meeting, if not exceeding, visitor expectations.

Table 12.1   Core principles of integrated quality management

1	Integration	Concern for quality, and the management techniques aimed at achieving it, should be integrated into all the tourism functions of the destination.
2	Authenticity	Visitors look for and appreciate genuine experiences. The quality of the tourism experience should not be devalued or fabricated.
3	Distinctiveness	Delivering quality should be concerned with bringing out the special, distinctive features and flavours of the destination.
4	Market realism	Quality management should be based on an informed and realistic assessment of the area's potential in the market place, identifying competitive strengths and ensuring they are not eroded.
5	Sustainability	Rural areas may often have fragile sites and small communities sensitive to intrusion and congestion. However, visitors are increasingly looking for unspoiled environments – delivering quality must be concerned with managing the impacts of tourism.
6	Consumer orientation	Quality management is about getting close to the visitor, understanding their needs and establishing whether they are met.
7	Inclusiveness	Quality should not be delivered to only the few – a good experience should be provided for all visitors, especially those with special needs.
8	Attention to detail	Be creative, but also pay attention to detail – provide enough information, check on facilities, provide extra services.
9	Rationalisation	A small number of good initiatives are better than many poor ones. Activities that are under-resourced and of poor quality should cease, or be combined into stronger more sustainable product.
10	Partnership	Working together is right in principle and essential for success. Quality cannot be delivered as a solo project – tourism enterprises/organisations and community groups should work together.

11	Interdependence	Quality rural tourism depends on, and in turn supports, many other activities such as agriculture, craft industries, transport and local services.
12	Time	Improving quality takes time. Success depends on planning for steady, achievable progress year on year.
13	Commitment	Personal enthusiasm and commitment to achieving quality is essential. This needs to involve the wider community.
14	Accurate communication	Providing visitors with accurate and up to date information is key to matching expectations with reality and ensuring satisfaction.
15	Monitoring	Quality management is all about regular monitoring and evaluation of impacts on the visitor, enterprises, the environment and the local community.

Source: European Commission (1999b)

The principles of IQM take into account the totality of a rural destination's tourism system, spanning public, private and voluntary sectors. It is an initiative that relies heavily on community participation, from both tourism-receptive and tourism-hostile perspectives, allowing a form of integrated dialogue to develop amongst visitors, tourism stakeholders and interest groups, which in turn drives sustainable development initiatives.

Tourism and Rural Tourism in Wales

Wales' dependence on tourism is scarcely matched by any other country in Europe. Tourism spending from overnight and day visitors contributes more than £2 billion directly to the Welsh economy, equivalent to 7 per cent of gross domestic product (Wales Tourist Board, 2000). In contrast, agriculture and forestry account for 2.4 per cent and construction 5.3 per cent of GDP respectively. Tourism is a major contributor to the economy of many rural areas in Wales and is increasingly championed as a catalyst for economic, social, cultural and environmental change.

A recent study commissioned by the Wales Tourist Board, National Assembly for Wales, Welsh Development Agency and Countryside Council for Wales (Newidiem, 2000) estimates that total expenditure by tourists in the rural areas of Wales is of the order of £1bn, which represents 50 per cent of total tourism revenue to the Principality. Tourism supports up to 100,000 jobs directly and indirectly in the Welsh economy, more than 10 per cent of the workforce (Wales Tourist Board, 2000). It is estimated that tourism supports 25,000 direct jobs in rural Wales (Wales Rural Forum, 1998).

The Study Area

Ceredigion is a rural county on the west coast of Wales, with a coastline fronting Cardigan Bay. With a population of approximately 70,000, its principal industries are agriculture and tourism, the latter accounting for 15 per cent of total employment in the county. Tourism is very important to the Ceredigion economy (Ceredigion County Council, 2001). In 1999, tourism spending in the county amounted to £115mn, of which 75 per cent was contributed by staying visitors and a quarter by day visitors. Accommodation attracts £26mn of this expenditure (23 per cent), the catering sector £36mn (31 per cent), retailing £24mn (21 per cent), attractions and entertainment £9mn (8 per cent) and travel £20mn (17 per cent). Some 4,372 actual jobs, or 3,137 full-time equivalent (FTE) jobs, are supported directly or indirectly by tourism.

The county council lists the following objectives for tourism development in Ceredigion (Ceredigion County Council, 2001):

1   to encourage, and where possible, to assist tourist operators to maintain, improve and enhance tourist attractions, particularly in respect of all-year-round facilities;
2   to promote sustainability and encourage developments which promote 'green' and 'cultural' tourism;
3   to maintain and improve public access within the countryside and improve the tourism infrastructure;
4   to reduce the over-dependence on the caravanning/camping sector and the impact of tourist caravans in the landscape and to broaden the accommodation base to provide for new markets, in the short breaks, overseas and conference/meetings sector.

Although its coastline and rural hinterland combine to offer the visitor an attractive holiday destination, Ceredigion, like many other areas in geographically marginal and peripheral regions, suffers from a number of problems in terms of tourism development. For example, it is a little known destination, with a low level of awareness outside of Wales itself. Tourist activity remains highly seasonal, with a marked peak in July and August. There is a heavy reliance on static caravans in the county's accommodation stock, at the expense of other serviced accommodation of high quality. In certain parts of the county, tourism infrastructure is under-developed.

Given the above weaknesses, while accepting the county's undoubted tourism potential especially in the growing activity holidays market, the county council decided, in conjunction with staff from the University of Wales in Aberystwyth, to initiate a project to investigate integrated quality management in rural tourism in the county. The project began in early 2002 for an initial 15-month period.

Project Aims and Methodology

The aim of the project is to assist the development of methods and systems to deliver and monitor integrated quality management in rural tourism destinations within Ceredigion as a way of contributing to the economic development of the county. Specifically, the project:

1   introduces the concept of integrated quality management to 'pilot' areas in the county identified through open competition;
2   provides assistance and support in the formulation and integration of the IQM principles within the selected pilot areas;
3   will monitor and evaluate the progress of the pilot groups;
4   will disseminate the project outcomes to the wider tourism community, thereby encouraging the adoption of IQM principles and practices throughout the county, elsewhere in Wales and further afield; and
5   will promote partnership and the exchange of best practice based on the IQM project evaluation and incorporating IQM guidance.

Central to the success of the project was the appointment of a project officer who is responsible for liaison with communities in the selected pilot areas. She has been required to introduce the concept of integrated quality management to the local communities, gain their confidence and assist them in progressing their own tourism developments at local level. The project officer is located within the Institute of Rural Studies at the University of Wales in Aberystwyth and has access to specialist staff and other resources to carry out the specified tasks.

The project officer works with communities in the pilot areas on three progressive strands:

1 Working together to a strategy:

   a)   setting the process going;
   b)   establishing leadership and partnership structures; and
   c)   developing a clear strategy, well communicated.

2 Delivering quality at all stages of the visitor experience:

   a)   marketing and communication;
   b)   welcome, orientation and information;
   c)   accommodation;
   d)   local produce and gastronomy;
   e)   attractions and events;
   f)   countryside recreation; and
   g)   environment and infrastructure.

3   Installing effective quality management and monitoring processes:

   a)   understanding visitor needs and seeing they are met;
   b)   setting, checking and communicating standards;
   c)   developing training and business support mechanisms; and
   d)   monitoring impacts on the local economy, community and environment.

Work Programme

The IQM project consists of three inter-related phases:

1   Planning phase

   Key tasks

   a)   familiarisation with the IQM concept through a review of existing literature and evaluation of examples of good practice;
   b)   familiarisation with the selected pilot areas in terms of existing tourism activity, key individuals, tourism statistics, tourist facilities, etc.; and
   c)   initial meetings with key individuals in pilot areas to explain the rationale and potential benefits of adopting IQM principles.

2   Pilot phase

   Key tasks

   a)   extensive liaison with the selected pilot communities;
   b)   development of a tourism strategy for each pilot area based on the IQM concept and principles, to include:
      i)   marketing and communication;
      ii)   welcome, orientation and information;
      iii)   accommodation;
      iv)   local produce and gastronomy;
      v)   attractions and events;
      vi)   countryside recreation; and
      vii)   environment and infrastructure.
   c)   introduction of the IQM concept and principles to the local communities via a series of meetings;
   d)   establishing leadership and partnership structures in the pilot areas;
   e)   identifying any training, research and information needs of the pilot areas;
   f)   identifying appropriate mechanisms for the adoption, ownership and development of the tourism strategy after the pilot phase; and
   g)   installing effective quality management and monitoring processes in line with IQM principles.

3   Pilot evaluation phase

Key tasks

a)   undertaking a comparative study to ascertain the strengths and weaknesses of the pilot projects;

b)   using the findings of the comparative study to develop a 'tool kit' that could be used to introduce the concept of IQM in rural tourism to other communities in Ceredigion and further afield; and

c)   disseminating the 'tool kit' to other communities via a series of open events, meetings and PR activities.

Progress to Date

At the time of writing, the IQM project is mid-way through its pilot phase. Communities within the three pilot areas have engaged with the project with varying degrees of enthusiasm and commitment. The pilot area that has been the most committed to embedding IQM principles has made extensive use of the project officer's knowledge and expertise to expand tourism activities in its community. The area that has been the most resistant to embracing IQM has been slow to take advantage of services offered, although there is cause to hope that this may change. Within these two extremes, the remaining pilot area now shows a commitment to IQM principles after a somewhat hesitant start and has been active in accessing capital funding for tourism developments from the Wales Tourist Board.

The project is proving very valuable in providing the pilot areas with detailed information on levels of visitor satisfaction with a range of tourism facilities, collected at local level via regular face-to-face and self-completed interview surveys. This type of detailed data has, hitherto, not been available at such a local level. Audits of existing tourism infrastructure, facilities and events have also helped to identify gaps in provision and have been instrumental in helping the appropriate local government authorities to prioritise their tourism-related developments. Taken together, the results of the audits and the surveys is being used as the basis for the development of detailed action plans for the pilot areas.

Future Outcomes of the Research

As discussed at an earlier point in this chapter, a review of the literature has revealed that mechanisms to deliver quality in rural tourism in an *integrated* manner are not well developed. Establishing integrated quality management (IQM) systems in rural tourism destinations offers an opportunity to forge a new direction for rural tourism in the twenty-first century. As it progresses, the IQM project in Ceredigion will seek to address a number of issues, including:

1   identifying obstacles to implementing IQM in rural tourist destinations;
2   devising strategies to overcome these obstacles;
3   harnessing community goodwill and expertise in strategy development;
4   establishing appropriate mechanisms for the adoption, ownership and development of the tourism strategy after the initial pilot phase of the project;
5   creating effective tourism and rural development partnerships;
6   collecting data on visitors and their requirements;
7   monitoring impacts of tourism on the local economy, community and the environment;
8   identifying the essential elements of the IQM rural tourism 'tool kit' as a means of disseminating good practice; and
9   identifying training, research and information needs of communities in tourism development areas.

The project officer is responsible for monitoring the progress of the project at local community level, ensuring that agreed targets are achieved. A steering group has been established to guide the direction of the project, thereby injecting expertise in relevant areas and ensuring the widest benefits to the community and county.

References

Augustyn, M. and Ho, S.K. (1998), 'Service Quality and Tourism', *Journal of Travel Research*, vol. 37, pp. 71-5.
Bramwell, B. and Lane, B. (2000), 'Collaboration and Partnerships in Tourism Planning', in B. Bramwell, and B. Lane, (eds), *Tourism Collaboration and Partnerships*, Channel View Publications, Clevedon.
Busby, G. and Rendle, S. (2000), 'The Transition from Tourism on Farms to Farm Tourism', *Tourism Management*, vol. 21, pp. 635-42.
Ceredigion County Council (2001), *Unitary Development Plan (pre-deposit version)*, Ceredigion County Council, Aberaeron.
Department for Culture, Media and Sport (DCMS) (1999), *Tomorrow's Tourism*, DCMS, London.
Diaz, A.M., Vazquez, R. and Ruiz, A.V. (1998), 'Factors Affecting Customer Evaluation of Service Quality in the Tourism Industry', in *Proceedings of the Annual Conference – European Marketing Academy*, Stockholm, 20-23 May.
English Tourism Council (2001a), *Time for Action: A Strategy for Sustainable Tourism in England*, English Tourism Council, London.
English Tourism Council (2001b), *Working for the Countryside: A Strategy for Rural Tourism in England 2001-2005*, English Tourism Council, London.
European Commission (1999a), *Towards Quality Coastal Tourism: Integrated Quality Management (IQM) of Coastal Tourist Destinations*, European Commission, Brussels.
European Commission (1999b), *Towards Quality Rural Tourism: Integrated Quality Management (IQM) of Rural Tourist Destinations*, European Commission, Brussels.
European Commission (1999c), *Towards Quality Urban Tourism: Integrated Quality Management (IQM) of Urban Tourist Destinations*, European Commission, Brussels.

Go, F.M. and Govers, R. (2000), 'Integrated Quality Management for Tourist Destinations: A European Perspective on Achieving Competitiveness', *Tourism Management*, vol. 21, pp. 79-88.

Lane, B. (1994), 'What is Rural Tourism?', *Journal of Sustainable Tourism*, vol. 2, pp. 7-21.

Long, P. and Lane, B. (2000), 'Rural Tourism Development', in W.C. Gartner and D.W. Lime (eds), *Trends in Outdoor Recreation, Leisure and Tourism*, CABI Publishing, Wallingford, pp. 299-308.

Newidiem (2000), *Farm and Agri-tourism Scoping Study*, Newidiem Consultants, Cardiff.

O'Neill, M. (2000), 'Service Quality: The Role of Perception in Disconfirmation Models of Service Quality in the Tourism Industry', *Measuring Business Excellence*, vol. 4, pp. 46-59.

Page, S.J. and Getz, D. (1997), *Rural Tourism: International Perspectives*, International Thomson Business Press, London.

Ritchie, B.J.R. and Crouch, G.I. (1997), 'Quality, Price and the Tourism Experience: Roles and Contribution to Destination Competitiveness', in P. Keller (ed.), *Reports of the 47th AIEST Congress*, Cha-Am, Thailand, pp. 117-39.

Wales Rural Forum (1998), *Rural Tourism: Towards Sustainability*, Wales Rural Forum, Carmarthen, Wales.

Wales Tourist Board (2000), *Achieving our Potential: A Tourism Strategy for Wales*, Wales Tourist Board, Cardiff.

Youell, R. (2001), *Foot and Mouth Disease: Impact on the Tourism Industry in Rural Wales*, IRS Briefing Paper 0301, Institute of Rural Studies, University of Wales, Aberystwyth.

Chapter 13

The Role of Education in the Management of Rural Tourism and Leisure

Patricija Verbole

Introduction

Quality of service delivery and customer satisfaction are critical tourism management challenges. Consumers' spending on services, such as holidays, recreation and leisure activities is increasing, as are customers' expectations with regard to quality, especially in the personalisation of services performed. Service quality and customer satisfaction are recognised as the primary factors affecting both repeat visiting and word-of-mouth promotion. Human resources therefore underpin the quality of the tourist product offered, and consequently of the customer satisfaction that further results in better self-promotion of the tourist destination, site, facility or activity. From this perspective, education and training of the professionals and families working in rural tourism and leisure play a major role in achieving sustainable tourism development. In this chapter, communication skills, as well as personal development skills, are seen as playing an important role in establishing good relations with customers and thus improving the quality of experience that can turn 'first time buyers' into 'regulars'. Consequently, this can result in better self-promotion of tourist products and services.

In this chapter, the term 'rural tourism' refers to all different forms of tourism found in rural areas, including agritourism and farm tourism, and to a number of special interest holidays and other contemporary leisure activities in the countryside, such as hiking, (mountain) biking, horseback riding, water and other sports, recreational enterprises and attractions. Due to increasing tourism demand, various new ventures are beginning to compete with farms for available resources and the income that may be derived from their use (Verbole, 2000).

Depending on the nature of the service offered, 'customers' are called guests, visitors, clients, users, tourists etc. For the purposes of this chapter the term customer is used to describe people who travel away from their normal place of residence for work, leisure, or to visit friends and family in the countryside, including day visitors and those who visit for 24 hours or more. Customers are a heterogeneous group of dynamic and often highly irrational beings whose motives, expectations, wants and needs are constantly changing. They expect increasingly high levels of quality, especially in terms of the personalisation of service performed (Torkildsen, 1993; Evans *et al.*, 1996; Perdue, 2000). If treated rudely at reception or unable to get a hot shower, customers can take their custom elsewhere. Recently, the disappearance of the loyal customer has been recognised as a worldwide phenomenon (Peters and Weiermair, 2000), due to a range of changes in consumer behaviour. Customers, irrespective of relative price and quality do not adhere to traditional, habitual, and repetitive purchasing decisions and behaviour. Prompted by more and easily accessible purchasing information, a greater number of new products and services and their fast global market diffusion, discriminating customers have become fickle, and disloyal to their traditional service providers.

The rural tourism or leisure product is something of a mystery. It is volatile, intangible, perishable, fragile, fleeting, and in many instances dependent on the person providing the service, such as the host (farm) family, tour guide, doorman, or waiter. Leisure participation is intangible until customers experience it. The product, namely the experience, cannot be stored, and is unpredictable. Managers, small-business owners, and farmers may each see the 'product' differently. It may be thought that the products are facilities (farmhouse) or activities (hiking). Yet, in reality these are the vehicles for achieving the real product (Otto and Ritchie, 2000). For example, a leisure activity product, that is experiencing satisfaction through leisure participation, is the unit of exchange with customers. If the customers do experience satisfaction, they will want to buy again. Hence, rural tourism and leisure providers' products are not merely comprised of facilities and activities but also of ways in which they are offered and the relationships created with the customers. Particularly in the context of rural tourism, product strengths are seen as personal contact, authenticity, heritage and individualism (Long and Lane, 2000).

Tourist enterprises (private and public) and family farms need to attract, create and keep customers both with products (e.g. horseback riding) and services (e.g. the ambience, friendliness) that customers seek and at the price they are prepared to pay, or ventures fail. Although good products to choose from are essential, intangible delivery services are becoming more and more important in achieving customer satisfaction. Torkildsen (1993: 4.10) cites care, attention and understanding, friendliness, helpfulness and cleanliness as factors that make customers feel welcome and important, and to be of significance in achieving customer satisfaction. To attract customers and increase their satisfaction management may consider adding or changing facilities, activities, services, creating new images, introducing new sports, organising new events, improving catering services, providing a wider range of thematic hiking routes, staying open longer hours, making an hotel foyer more attractive, improving added customer services, or training staff in customer care. In the end, it is all about selling benefits; customers want to enjoy, to be with friends, to learn, to rest, to feel better, to look better, to be skilful, or to win. In the latter case, for example, it is necessary to sell 'success'. Management has the means to make people happier, healthier, fitter, to make them better players, to make them feel glamorous, be risk takers, feel excited and be adventurous. In the tourism industry providers are commonly concerned with the tangible aspects of tourism development, in particular in its first phase, for example the creation of new hotels and traditional infrastructure, and less with the creation of intangible benefits in terms of 'tourism experiences' (Peters and Weiermair, 2000). In cases of outdated and declining rural tourism and leisure products, intangible service characteristics may provide the competitive advantage. Increased attractiveness of destinations, sites, facilities, or activities, can therefore be achieved through appropriate management.

Customer-oriented Management

Rural tourism and leisure, regardless of who manages them, are market-orientated businesses – their provision consists of infrastructure and facilities, leisure programmes, products that are priced, promoted and targeted, and the performance being evaluated. From a strategic point of view, a main objective of tourism and leisure management is to increase the probability and frequency of consumers coming into contact with the product and services, purchasing and using them, and subsequently repurchasing them; that is to gain customers who will return and promote the products and services offered to their friends and relatives. Although rural tourism and leisure management may have different goals to achieve in different settings, they should always be profit-orientated. This profit, however, can be measured not solely in terms of money, but by many success criteria. Customer satisfaction is one key criterion (Evans *et al.*, 1996; Witt and Muhlemann, 2000). It is generally argued that, if customers are satisfied with the product or service, they are more likely to return and to tell others of their favourable experiences (Moscardo, 1999). Management that is focused on customers, objectives and continuous improvement, is thus a prerequisite for good customer relations and customer care.

The role of management here is two-fold. First, customers evaluate how they are treated – and they want to be treated with respect, understanding, attention and importance – which requires enterprises to focus on them. Customer care, however, is about more than relating pleasantly to customers; it is linked to the achievement of the organisation's strategic objectives. Second, management should also consider employees or family members as internal customers. To enhance customers' experience, it is essential that people working in rural tourism and leisure have access to the relevant education, knowledge and skills to enable them to communicate effectively with customers and others involved in all stages of service delivery. For example, it is often the personnel at the 'sharp end' that are the least capable of warm communication with customers, yet who are called upon to undertake the most important job – that of communicating with people. Employees or family members need help in carrying out these important functions; they need training. Regrettably, many rural tourism and leisure providers, far from motivating people, often serve to demotivate them, achieving the opposite of that which was intended. Therefore, management further needs to find ways to motivate and train personnel for quality service performance, for example by encouraging active participation in decision making, delegating responsibilities, and focusing on results rather than on tasks. A major cornerstone of achieving competitive advantage is therefore concern for customer satisfaction with services delivered, and the finding of ways to increase employee performance by identifying more enjoyable ways of doing the job (Perdue, 2000).

Enhancing Service Delivery Process

The most successful service delivery processes are the ones that are 'orchestrated, planned and processed like a screenplay with a well written script' (Peters and Weiermair, 2000: 27).   Torkildsen (1993: 4.11) looks at the process of service delivery as a chain of many 'events' or 'interactions' between customer and service provider, that can be clearly visualised by getting the service mapped out onto a flowchart. At each link in the chain satisfactory service is required, and good customer care can be considered only as good as 'the weakest link in the customer care chain'. The customer's impression of a service or facility is often made on the flimsiest brush or smallest incident with a member of personnel or family.

Service providers and hosts should realise that several discrepancies between their own views and customers' expectations and perceptions, may occur in the process of delivering their services. Based on personal needs, prior experiences, word-of-mouth promotion and external marketing communication about service, customers have certain expectations about the quality of the product and services. Service providers, on the other hand, have their own ideas about customers' expectations, which they translate into service quality guidelines and specifications. Based on an individual judgement and in accordance with the developed guidelines and specifications, the actual service is performed, and customers' perceptions of the quality and other attributes of the actual service are formed as a result. Even though the service provider may perceive a job done as excellent, the customer may evaluate the service differently. As customers compare the perceived service quality with the expected one, they may be satisfied if the quality of service performed was at least as good as expected, or dissatisfied if the performance was worse than expected. The expectations may be lowered or raised during the service process, as the customer becomes more experienced.

Any gaps that appear (in customer expectations, service product design, actual service delivery or marketing communication or example) are the concern of the providers, and to stay 'in business' such gaps should be made as small as possible. As customers are involved not only in transactions, but also in *in situ* production of the service, effective communication can be seen as the key element in improvement of service quality and provision of a quality experience for customers (Moscardo, 1999). In general, whoever manages must be aware that skilled employees and family members form the basis of customer satisfaction. In terms of rural tourism and leisure, awareness of this can even be considered more crucial due to the nature of the customer-host relationship. Rural tourism allows for close personal contact with the host. This pull-factor, together with hospitality, has been recognised as a competitive advantage in itself (Long and Lane, 2000), despite the fact that it might also be seen as a challenge for rural entrepreneurs, marketers and service providers. Understanding customers' perceptions, for example, may be more challenging in rural tourism and leisure industries, due to considerable differences that might exist in the backgrounds and worldviews of customers and hosts.

Customers and personnel should actively participate in service and delivery design and specifications development, that are usually seen as the preserve of management as their knowledge and ideas can make a valuable contribution in negotiating and formulating shared objectives. As Peters and Weiermair (2000) stress, it is important to integrate the customer intellectually and emotionally into the service production process, and not merely provide the physical integration of the customer with the service. An active participation and the collective development of any action plan immediately ensures supported implementation, as everyone has contributed and therefore better understands how and why decisions were made.

This introduces the idea that management and personnel should work together as a team to achieve the best performance. It is very important to have the 'best people for the job' with abilities, experiences, qualifications and skill within the field of rural tourism and leisure as elsewhere. Nevertheless, people's ability to behave in certain ways, and to contribute and inter-relate are shaped by their personality and learned behaviour, and these are also seen as important assets in their teamwork performance (Knight, 1995; Witt and Muhlemann, 2000). To apply the idea of the teamwork to the case of farm or agritourism, gender and inter-generational relations at the rural household level would need to be taken into consideration.

Delivering services offers many opportunities and threats in terms of communication. To benefit from it, people involved in delivering services should have communication skills (Littlejohn, 1996; Cole, 1993; Moscardo, 1999) to get to know the customers, their needs, wants and expectations, and consequently improve service quality by matching or satisfying these (Peters and Weiermair, 2000; Perdue, 2000; Evans *et al.*, 1996). To be able to provide a good service and thus, a quality customer experience, it is first of all necessary to get to know the customers, find out their needs, wants and wishes, as well as collect data on customer characteristics, such as social group, culture, knowledge and familiarity with the rural setting, for example. Within an heterogeneous group, different customers may have many similar needs. Likewise, similar types of customer may have different needs influencing a selection from many choices available in rural tourism and leisure. Management needs to be aware, for example, that customers can also decide or form their responses based on various demotivators rather than on what motivates them. As the greatest demotivators, Torkildsen (1993) identified poor handling of customers (causing delays and queues), rudeness, abruptness and take-it-or-leave-it attitudes, ruined expectations, dissatisfaction or poor service, broken promises, double-bookings, as well as tangible issues such as dirty toilets, or a lack of hot water. Thus, it is important that the perceptions of potential customers and their expectations are met when coming in personal contact with service providers.

If they are aware of the importance of such issues, small-scale rural enterprises that invite more personal contact with customers, can be considered to have an advantage over large organisations in this respect. There are various methods that can be used to gather information on customers (see Cole, 1993), such as observing, listening, asking and monitoring. All employees, particularly those who have more direct customer contact, like doormen, receptionists, and catering or bar personnel, can be receptive to clues and signs of happiness or of displeasure. This is one of the keys to successful customer relations, as it influences customers' responses to the providers' communication efforts (Moscardo, 1999).

The discrepancy between demand and supply side perceptions may be therefore reduced through the dynamic communication process of exchanging views, opinions, ideas, and even complaints. This refers not only to the communication with customers, but also employees and family members, who can be considered as 'internal' customers. Thus every customer has to be treated with respect, importance, attention, appreciation and understanding. This underlines the importance of training for people working in tourism so that they understand the process of managing complex service chains. Small and medium-sized tourism enterprises can learn from the managerial know-how of successful entrepreneurs. As a minimum, service providers should understand customer problems and behaviour in order to fulfil customers' expectations; preferably they should aim to provide them with an extraordinary service experience.

Tourism and Hospitality Education and Training

Educated, skilled, knowledgeable and motivated people involved in the process of service delivery play an important role in effectiveness and efficiency of any tourism and hospitality business, including those in rural tourism and leisure. Positive attitudes to training and life-long learning among employees and family members should be encouraged, as they lie at the core of the total quality concept. People working in rural tourism and leisure – leaders, facilitators, trainers, managers, and families – need a wide variety of knowledge and skills (Cooper *et al.*, 1998) that can be obtained through formal and informal education and training. The possession of certain skills, however, is not enough; personal commitment and culture change make a difference (Witt and Muhlemann, 2000). Education and training should allow for the acquisition of multidisciplinary knowledge, which gives management and personnel the necessary flexibility and creativity required in these fast changing times. Specialised knowledge is too often 'a trap for creativity' (Löhmöller, 1997: 39). Education and training should also ensure participants' action on the ground, and should be tailored to the needs of the people and rural areas. In addition to business skills, rural tourism development, micro- and small-scale rural tourism and leisure enterprise, anthropology, psychology and sociology, themes such as the following should be covered:

- personal effectiveness and collaboration (in terms of personal development and change, learning styles, team building, conflict handling);
- communication (multicultural, interpersonal, foreign languages, verbal and non-verbal);
- management (in terms of decision making, problem analysis and solving);
- leadership and human resource management;
- consumer behaviour; and public relations; and
- visitor management.

The satisfaction of clearly identified needs for such education and training in the management for rural leisure and tourism seems to be difficult. Research conducted by McKercher and Robbins (1998), for example, highlights lack of training and business planning as a major problem in small rural tourism businesses due to limited time, finances and personnel. Similarly, Verbole (2002a, 2002b), in a study conducted among micro- and small-sized enterprises in mountainous tourist regions in Slovenia, reports that in many cases entrepreneurs have been pushed towards their entrepreneurial activities, lacking formal business education and experiences. Aware of this 'handicap', and also having lower levels of formal education, rarely in tourism-related fields, rural tourism and leisure entrepreneurs are eager to learn. Unfortunately, existing capacity-building and training opportunities are limited; as is awareness of them. Many opportunities that are available are found irrelevant or insufficiently tailored to the needs of entrepreneurs who lack time, finance, mobility and personnel to attend courses or visit business support centres in distant urban places. Research (Kampuš Trop, 2000; Birch, 1993; Vahčič, 1995; Florjančič and Jesenko, 1997) highlights that support activities, educational provision, communication and transport infrastructure are unequally developed across urban and rural environments, resulting in concentration of ventures in urban environments with well-developed infrastructure and good living conditions. There are calls for proper forms of education and training (Löhmöller, 1997; Verbole, 2002b), better tailored to the needs of rural tourism and leisure entrepreneurs in terms of contents, techniques and methods, duration, time, location and costs.

Formal education curricula need to be designed to teach students technical and other skills previously suggested that will help them to become successful multiskilled professionals in rural tourism and leisure. In line with Verbole's (1999) identification of a primary school as one of the important key actors in the process of rural tourism and leisure development, formal education can be seen as valuable support in enabling students to develop certain attitudes, norms, and gain knowledge and skills that would enable them to actively participate in and benefit from tourism. At all levels, topics relevant to tourism should be integrated in the curriculum of every subject, including outside environments, providing after-school activities and interest clubs. This will result in the learning of communication skills, foreign languages, different cultures, places and people, behavioural codes and standards, and about the importance of natural and cultural heritage preservation.

In Slovenia for example, tourism managers who graduated mostly in economics, law or management, find tourism and hospitality vocational schools and college programmes suitable, whereas the undergraduate courses are poorly developed (Jereb, 1997) especially with regards to (rural) tourism and leisure. An emphasis is put on management in larger organisations rather than in the small-sized enterprises common to rural areas, and on economics, law, and technical skills, neglecting the role of personnel management and total quality assurance as well as communication and personal development skills. Similarly, subjects taught at universities in tourism-related fields, such as economics, social sciences and agriculture, lack the tourism perspective. The Slovenian formal education system fails to equip its graduates with the knowledge and skills necessary to work effectively in the complexities of rural tourism and leisure industries. Education is narrowed down to specific fields of interest which do not address development issues from a broader perspective. Worldwide, however, formal educational institutions can be found that are committed to equipping students with diverse and appropriate skills to enable them to meet rural tourists' or leisure participants' ever increasing demands (for example The Scottish Agricultural College, Auchincruive, and Wageningen University and Research Centre).

Further ways to bring relevant knowledge and skills closer to people working rural tourism and leisure are provided by various forms of informal education and training, and through many associations and institutes for adult education as well as official and private consulting organisations (Petrin, 1997; Verbole, 2002b). Training can take the form of traditional seminars and courses aimed at optimising the service delivery processes, the creation of new products and services and enhanced service delivery processes, as well as at stabilising the enterprise by positive changes of personnel. All such education and training aims to offer a wide basis of knowledge for more creativity and flexibility in order to ensure ventures' survival in the future (Löhmöller, 1997).

In addition, workshops should be considered as a convenient form of training, in which the 'learning by doing' process and 'learning by experience' is central to participants' development. The following should be considered when organising workshops, and included as appropriate:

- individual exercises and role playing;
- group exercises for problem solving and decision making;
- focus/key questions;
- brain maps;
- problem trees;
- case studies;
- lecture inputs; and
- plenary discussions, information papers.

Rural communities should foster different institutions and a variety of partnerships to support local development. One unique approach to rural development is the business incubator (Petrin, 1997). Amongst other things, the incubator can be designed to encourage people in rural areas to start a venture, to promote specific types of businesses, to help develop networks, and to intensify training programmes to build the vocational skills of incubator members. It improves access to capital, and can provide jobs for graduates. Similarly, small business support centres can be established to meet the needs of start-ups and enable networking. Experience shows that personal and organisational networks are found to be very effective (Petrin, 1997). As reported by Verbole (2002a), among rural tourism and leisure entrepreneurs, informal social gatherings and events are seen as excellent opportunities to exchange experiences, ideas and views.

In order to satisfy the specific needs of rural tourism and leisure entrepreneurs and families, personal counselling and advice can prove most valuable, allowing for *in situ* training and assistance, as well as continuous information dissemination (Verbole, 2002b). Consultants need to become partners in communication, able to grasp and understand the thinking processes of their customers so as to provide a high quality service. It is important that counselling support, sometimes given through business support centres, manages to cover a designated area and relate to its specific characteristics. Where support is generalised and prioritises the internationalisation of enterprises, company networking and cluster development, increasing productivity, technological development and innovation, rural tourism and leisure entrepreneurs may be further marginalised by, rather than drawn into, development processes. Finding advisers and trainers to facilitate appropriate training, knowledge and skills proves to be critical in the development of rural tourism and leisure entrepreneurship (van den Ban and Hawkins, 1996; Knight, 1995).

Finally, mass media, such as television, radio, videos, newspapers and newsletters, can play an important role in education for successful rural tourism and leisure development and should not be overlooked. Mass media can be used as a powerful tool for spreading information to the wider population, local people, and especially to farm families in more remote rural areas, as well as creating an entrepreneurial and customer-oriented culture (Rice and Atkin, 1989; Behrens and Evans, 1984; Verbole, 2002c).

Conclusions and Discussion

In the intense competition of dynamic contemporary markets offering a wide choice of products and services, quality is essential. In order to turn first time purchasers into repeat customers, a key business aim, the customer's experience and satisfaction are paramount and require a high quality service delivery. Sometimes it is necessary to change people in order to change things. Success therefore depends on the interests, abilities and capacities of people to bring about change. Educated, knowledgeable, multi-skilled and motivated people working in rural tourism and leisure, have an important role to play in inculcating good relationships, and are as such the core of the total quality concept. To allow for effective communication with external and internal customers, training is needed in personal development, in communication, management, leadership and interpersonal skills, and in decision making, problem analysis and solving, and teambuilding. Both formal and informal training should be tailored to the specific needs of rural tourism and leisure entrepreneurs and families in terms of content, techniques and methods, duration, time, location and costs. *In situ* advice and counselling, informal social gatherings, workshops, and mass media prove to be the most effective options.

References

Behrens, J.H. and Evans, J.F. (1984), 'Using Mass Media for Extension Teaching', in S.S.E. Burton (ed.), *Agricultural Extension: A Reference Manual*, FAO, Rome, pp. 144-55.

Birch, D. (1993), 'Dynamic Entrepreneurship and Job Creation: Lessons from the US Experience for Central and Eastern Europe', in D.F. Abell and T. Köllermeir (eds), *Dynamic Entrepreneurship in Central and Eastern Europe*, Delwel, The Hague.

Cole, K. (1993), *Crystal Clear Communication: Skills for Understanding and Being Understood*, Prentice Hall, Sydney.

Cooper, C., Fletcher, J., Gilbert, D., Shepherd, R. and Wanhill, S. (eds) (1998), *Tourism: Principles and Practice*, Longman, Harlow, 2nd edn.

Evans, M.J., Moutinho, L. and Raaij, W.F. van (1996), *Applied Consumer Behaviour*, Addison-Wesley, Harlow.

Florjančič, J. and Jesenko, J. (eds) (1997), *Management v Turizmu*, Moderna organizacija, Kranj.

Gartner, W.C. and Lime, D.W. (eds) (2000), *Trends in Outdoor Recreation, Leisure and Tourism*, CABI Publishing, Wallingford.

Jereb, J. (1997), 'Kadrovski Management v Turistični Dejavnosti', in J. Florjančič and J. Jesenko (eds), *Management v Turizmu*, Moderna organizacija, Kranj, pp. 63-85.

Kampuš Trop, V. (2000), *Podjetništvo v Sloveniji – Regionalne Razlike v Ustanavljanju in Razvoju Novih Podjetij*, unpublished PhD thesis, University of Ljubljana, Ljubljana.

Knight, S. (1995), *NLP at Work: The Difference that Makes a Difference in Business*, Nicholas Brealey, London.

Littlejohn, S.W. (1996), *Theories of Human Communication*, Wadsworth Publishing, Belmont, 5th edn.

Long, P. and Lane, B. (2000), 'Rural Tourism Development', in W.C. Gartner and D.W. Lime (eds), *Trends in Outdoor Recreation, Leisure and Tourism*, CABI Publishing, Wallingford, pp. 299-308.

Löhmöller, G. (1997), 'Concept for the Development of Entrepreneurial Activities in the Rural Area for Farmers and Managers of Small- and Medium-sized Enterprises', in T. Petrin and A. Gannon (eds), *Rural Development Through Entrepreneurship*, FAO, Rome, pp. 31-43.

McKercher, B. and Robbins, B. (1998), 'Business Development Issues Affecting Nature-based Tourism Operators in Australia, *Journal of Sustainable Tourism*, vol. 6, pp. 173-88.

Moscardo, G. (1999), *Making Visitors Mindful: Principles for Creating Quality Sustainable Visitor Experience Through Effective Communication*, Sagamore Publishing, Champaign, Illinois.

Otto, J.E. and Ritchie, J.R.B. (2000), 'The Service Experience in Tourism', in C. Ryan and S. Page (eds), *Tourism Management: Towards the New Millennium*, Pergamon, Amsterdam, pp. 404-19.

Perdue, R.R. (2000), 'Service Quality in Resort Settings: Trends in the Application in Information Technology', in W.C. Gartner and D.W. Lime (eds), *Trends in Outdoor Recreation, Leisure and Tourism*, CABI Publishing, Wallingford, pp. 357-64.

Peters, M. and Weiermair, K. (2000), 'Tourist Attractions and Attracted Tourists: How to Satisfy Today's "Fickle" Tourist Clientele?', *The Journal of Tourism Studies*, vol. 11, pp. 22-29.

Petrin, T. (1997), 'Institutions Supporting Entrepreneurial Restructuring of Rural Areas', in T. Petrin and A. Gannon (eds), *Rural Development through Entrepreneurship*, FAO, Rome, pp. 45-55.

Rice, R.E. and Atkin, C.K. (eds) (1989), *Public Communication Campaigns*, Sage, Newbury Park, 2nd edn.

Torkildsen, G. (1993), *Torkildsen's Guides to Leisure Management*, Longman, London.

Vahčič, A. (1995), *Some Regional Aspects of Entrepreneurship Economics in Slovenia*, Proceedings of the Workshop on Regional Aspects of Entrepreneurship in Eastern Europe, University of Tódz and University of Pittsburgh.

van den Ban, A.W. and Hawkins, H.S. (1996), *Agricultural Extension*, Blackwell Science, Carlton, 2nd edn.

Verbole, A. (1999), *Negotiating Rural Tourism Development at the Local Level: A Case Study in Pišece*, unpublished PhD thesis, Wageningen University, Wageningen.

Verbole, A. (2000), 'Actors, Discourses and Interfaces of Rural Tourism Development at the Local Community Level in Slovenia: Social and Political Dimensions of the Rural Tourism Development Process', *Journal of Sustainable Tourism*, vol. 8, pp. 479-90.

Verbole, P. (2002a), *Challenges of Being an Entrepreneur: Tells of Slovenian Mountain Women*, Opening Paper, International Preparatory Meeting Celebrating Mountain Women, EMF, Chambery, France, May, <http://www.mtnforum.org/europe/cmw/thinputs/03entrepr/entrepr01_pv.htm> .

Verbole, P. (2002b), *Mountain Women Entrepreneurship in Slovenia: Policies and Practices*, Paper presented at the International Conference Celebrating Mountain Women, ICIMOD, Paro-Thimphu, Bhutan, October.

Verbole, P. (2002c), *Communicating Environmentally Sustainable Alpine Tourist Behaviour through Print Tourism Promotion*, unpublished MSc. thesis, Wageningen University and Research Centre, Wageningen.

Witt, C.A. and Muhlemann, A.P. (2000), 'The Implementation of Total Quality Management in Tourism: Some Guidelines', in C. Ryan and S. Page (eds), *Tourism Management: Towards the New Millennium*, Pergamon, Amsterdam, pp. 390-441.

Chapter 14

Ecotourism for Rural Development in the Canary Islands and the Caribbean

Donald Macleod

Introduction

This chapter focuses on the phenomenon of ecotourism and looks at how it can develop in practice. It examines two case studies, both quite different, particularly in their respective creation and planning. Ecotourism is a term that has often been analysed and deconstructed, and it remains as an ideal and a driving force for many projects worldwide. With the concepts of sustainable development, community involvement and nature conservation included within its embrace, it is an admirable and potentially constructive idea. In practice, however, the reality of ecotourism is often different from the ideal. Drawing on anthropological research findings based on long-term fieldwork the chapter will describe the grass-roots reality of the projects in their rural settings. The comparison will allow the reader to see the relative importance of the cultural context of ecotourism, in particular the socio-economic and political environment in which the projects develop.

The first case study deals with La Gomera in the Canary Islands, where alternative tourism has dominated. Here the ecotourism has developed in a way that will be described as 'spontaneous' development. This means that there has been no organised, calculated planning by a specific group of professionals, but rather a response by local individuals and family groups, operating privately and making their own judgements and decisions. This is partly because of the predominant types of tourists visiting, as well as the character of the place itself. More recently, commercially minded entrepreneurs have capitalised on the natural assets and the type of tourists staying in the valley. They have introduced 'bike tours, 'dolphin tours' and guided mountain walks and have stimulated blatant commercial ecotourism for package tourists at the new hotel and apartment blocks.

In contrast, the chapter also looks at the Dominican Republic, a place where mass tourism is predominant, where tourists generally stay in large hotels (500 beds) within compounds, enjoying a protected beach, water sports and organised coach trips. Yet there are pockets of ecotourism in the country, notably whale-watching in the north east and mountain walking in the central uplands. There is also a move to establish a small ecotour into the National Park Del Este, a recognised 'Park in Peril' (Brandon et al., 1998). Situated in the south east coastal region, Del Este embraces an island 'Saona' to which over a thousand tourists are transported every day to experience the pristine beaches. Meanwhile the forested interior of the park remains rarely visited and a major project to introduce ecotourism into the park is underway. This chapter will look at the ideas, hopes and planning which have gone into the project, involving numerous public organisations and Non Governmental Organisation (NGOs), and it will highlight the problems of such an endeavour.

Ecotourism has been discussed as a term by numerous scholars and below are some definitions and assessments that reflect the interest and inclinations of this chapter.

Western (1993: 7) sees ecotourism's roots lying in nature and outdoor tourism, and in particular, 'specialist tours – birding safaris, camel treks, guided nature walks and so on… . Ecotourism is really an amalgam of interests arising out of environmental, economic, and social concerns'. He gives the official definition of the US Ecotourism Society, publishers of his book: 'Ecotourism is responsible travel to natural areas which conserves the environment and improves the welfare of local people' (Western, 1993: 8).

In more broadly human terms, Western (1993: 8) sees ecotourism as 'creating and satisfying a hunger for nature, exploiting tourism's potential for conservation and development and averting its negative impact on ecology, culture and aesthetics'. Importantly, he sees it as 'shifting from a definition of small-scale nature tourism to a set of principles applicable to any nature-related tourism' (Western, 1993: 10).

The World Tourism Organisation, in an article on the 'International Year of Ecotourism, 2002', claims that there is not a universal definition for ecotourism. It lists five general characteristics, number one of which is: 'All nature-based forms of tourism in which the main motivation of the tourists is the observation and appreciation of nature as well as the traditional cultures prevailing in the areas.' The remaining criteria mention education, small size, negative impact, and the protection of natural areas (WTO, 2001).

Finally, with a similar emphasis on nature tourism, Mieczkowski (1995: 460) regards ecotourism as a subset of 'alternative tourism' (which he opposes to 'mass tourism') – being 'small-scale, low-density, dispersed in non-urban areas, catering to special interest groups…'. He views ecotourism as being nature-oriented and nature-based, and does not wish to include cultural heritage within the definition.

The above definitions would embrace the examples this chapter examines, with the focus on an involvement with 'nature' on a small-scale and intimate level. However, the studies will reveal the importance of the receiving (host) human social environment as it impacts upon the ecotourism project in different settings. Overall, the chapter will illustrate how rural tourism, in the guise of ecotourism, has become an important and sometimes vital element in the development of rural areas around the world. It shows that a global issue, sustainability, has become embraced by rural communities experiencing tourism, and that the practical outcomes of abstract ideas and intentions are not always as anticipated. Such developments should be monitored if we are to learn from the actual grassroots experience of the communities and understand the direction in which rural tourism is heading and its impact.

Vueltas, La Gomera: Spontaneous Development

La Gomera is one of the smaller Canary Islands, situated near Tenerife. It is mountainous, with peaks reaching 1500m and possesses an ancient (two million years) temperate rainforest in its centre. This area has been designated a national park and is also a World Heritage Site. The region studied as part of an anthropological fieldwork project is in the south west, and part of a huge valley (Valle Gran Rey), which leads down to the sea. On one side of the coastal plane there lies an area of scrubland where a fishing settlement 'Vueltas' developed at the beginning of the 20th century.

Tourism began on La Gomera during the mid-1970s and was composed of explorative individuals moving away from the mass-market destinations of other islands, onto the totally unprepared La Gomera, whose transport infrastructure was poor – metalled roads and electricity being very recent additions in the west. These early visitors arrived in small numbers and were accommodated in the few bars/restaurants that offered rooms. Eventually, in the valley, two American settlers and a German settler opened up apartments for the growing tourist trade. This pattern continued and was boosted by the establishment of a direct car-ferry link between Los Cristianos on the south of Tenerife and San Sebastian, the main port of La Gomera.

By the early 1990s, the predominant visitors to the valley were German 'alternative' tourists (Macleod, 1993, 1997, 1998). These were independent travellers, often backpackers, seeking accommodation on arrival, and usually under 45 years of age. A cottage industry of family-based accommodation 'apartments' had developed. In the port of Vueltas, with its population of 350, some 36 sets of apartment buildings offering beds for 300 guests were constructed, mostly in the mid-1980s. One fisherman claims to have started the fashion for extending homes vertically to create tourist apartments, and most families were able to offer rooms for rent.

The tourists themselves were interested in the 'laid-back' holiday scene, visiting the different beaches, enjoying the bars and restaurants, taking trips around the island and walking in the mountains and forest. There was no tourism information centre in the valley from which to obtain information about facilities or excursions. There were no organised tours. People learnt from friends or by word of mouth about the island and its assets. Some walking guides had been published in German, and a general guide to the island also told of the natural wonders available. Equally, there were no maritime excursions – in contrast with Los Cristianos, Tenerife where dolphin and whale tours were well established.

Nevertheless, the tourists were enchanted with the natural beaches, the impressive valley with its palms, bamboos and riverine flora, as well as other local features including a walk to a waterfall and the spectacular mountain passes. Many spent the days body-surfing the powerful waves off an unspoilt beach, when occasional dolphins in the distance would cause a stir among the sunbathers. The overall ambience of the resort was one of unflustered relaxation, calm and contemplation – a place where individuals were free to unwind and enjoy the unmodified natural assets.

However, a number of developments during the 1990s accentuated the resorts' qualities and built on the type of tourist visiting the region. By 1991 some German entrepreneurs were busy selling images of the island in the form of postcards and books focusing on its natural charms and traditional folk culture. Another had started renting out mountain bikes and giving bike tours. Additionally, one had opened a maritime centre ('Club de Mar') selling fishing tackle and offering general advice to tourists. The maritime centre developed into a business offering sailing tours and fishing trips after 1991. By the late 1990s it had transformed into a dolphin and whale tour centre, giving daily trips hosted by experts, showing weekly films on the marine life and also serving as a research base.

Demand for the dolphin trips has remained high, reflecting the continued popularity of the resort, and since the mid-1990s two more organisations have begun offering 'dolphin safaris' on much larger cruise boats staffed by ex-fishermen from the islands. These boats actually travel around half the island sightseeing, with snorkelling and a locally caught fish lunch included. The chance of seeing dolphins is a bonus. The crews have a local knowledge of the area and of the sea creatures, and as fishermen theirs has been largely a predatory role. The tours are undoubtedly commercially driven operations. They contrast strongly with the German-run dolphin tours, in that the scale of the German operation is smaller, using 11m ex-fishing boats, with a focus on seeing cetaceans. Furthermore, the North European crew have a serious personal interest in the mammals, and could be described as ecologically motivated. The knowledge of the Maritime Centre staff is detailed, and records are kept of sightings and behaviour: more than seven dolphin and 11 whale species, including the Blue Whale, have been sighted, and a valuable data-base of movements in the region has been accumulated.

To put these marine-based ecotours in economic perspective, they could potentially generate more annual income (between the three tour companies) than the entire fishing fleet in Vueltas: around 40 professional fishermen with only 20 fishing full-time in 2002. Considering the dolphin tours have been running for less than ten years, and the fishermen established for over 90 years, this is a considerable feat. And yet, it is a pattern indicative of maritime communities worldwide, including our example from Dominican Republic, where global changes in technology, availability of fish, taste and demand have transformed the fishing industry, and tourism has continued to grow.

These developments on La Gomera were not part of a self-conscious, professional planning initiative, but rather, private, commercially-based phenomena, arising out of an intimate knowledge of the consumer market, which might be termed 'spontaneous development'. They should be seen against the background of major planned development initiatives in the region, sponsored by the European Union and the Spanish Government, which have sought to increase the economic and social wellbeing of the population. As such, 'structural funding' from the European Union Structural Development Fund (FEDER) has supported the building of roads, tunnels and an airport, among other major civil engineering projects. One early initiative, proposed in 1989, to 'redevelop' the beach by building groynes and adding sand to the seasonally denuded beach, was vigorously opposed by local people, ecological groups and others. It became a rallying point for protest against environmental destruction, and enjoyed the support of many tourists, and some signed petitions against the construction of environmentally destructive roads (Macleod, 1999). Nevertheless, during 1999 there were new plans afoot to rebuild the port at Vueltas enabling it to cater for a car ferry, which would dramatically transform the character of the village.

By the year 2000 there were private companies offering scuba-diving, mountain-biking, mountain-walking, fishing trips, whale and dolphin watching, boat tours and donkey-riding, all based in the valley. A hotel has been built at the seafront, and a tourism information office advises on excursions and apartments. The profile of visitors is changing, favouring those taking package holidays booked from their home countries and staying in the hotel or larger apartment complexes that now outnumber the private homes offering rooms. Many of these new package tourists go on planned excursions with guides employed by the hotels – and they also take the 'dolphin safari' tours.

The spontaneous development of ecotourism in the valley is becoming more business-like: commercial and competitive. It is clearly arguable that the increase in volume of tourists roaming pathways and taking boat-trips may pose a threat to the sensitive ecosystems and animals being appreciated. Yet in terms of economic productivity and environmental awareness, the ecotourism in the valley is a success, building on a latent market, in a rural area where the natural setting has been the main attraction.

National Park Del Este, Dominican Republic: Planning for Ecotourism

In contrast to La Gomera, the Dominican Republic has been oriented towards the mass tourism package market, dominated by large hotels based in secure complexes along the coast. It now offers very competitive holidays in the Caribbean and aims at the North American and European market. There are, however, pockets of small-scale rural tourism, and one of these at Bayahibe (population 1,800), situated on the edge of National Park Del Este on the south east coast, is the base for this case study.

Bayahibe had been a fishing village for most of the twentieth century although for some time in its past it was also the home of farmers and foresters. Recently it has experienced a transformation into a village whose economy is supported mainly by tourism. This is an indirect support, as the means of income is based on driving boats for hotel tourists arriving by coach from other parts of the island. Tourists are driven in fast launches to the small island of Saona, which is part of the National Park Del Este and has pristine beaches where they while away most of the day.

Fishing, which was the mainstay of the village economy, has now declined so that only 15 full-time fishers operated in the year 2000, although many other men would place traps in the sea, and some launch captains would fish during their working hours whilst waiting for tourists. In contrast to the scenario in Bayahibe, on the other side of the park the village of Boca de Yuma is brimful of fishermen (more than 200) and they harvest the waters around Saona. They claim there is no other form of employment in the locale. This situation has serious consequences for the national park and the regional ecology as the channel between Saona and the mainland is the nursery for many sea creatures and sustains the marine life of the south coast of the Dominican Republic. As a result of the continual sea traffic from Bayahibe to Saona, the fish have deserted the coast, and fishermen claim they must fish in the channel or just off the island of Saona to maintain their catch.

The park has restrictions on activities within its terrestrial boundaries. But there is no recognised jurisdiction over the channel and coastal waters – this is partly due to the fact that the navy polices the sea, and also because in drawing up the national park management jurisdiction, an omission was made regarding the channel. Within the park itself there are rules restricting the pasturing of animals, the hunting of animals, including birds and crabs, the collection of firewood and so on. The park is said to harbour the rare Hispaniolan parrot, the Crowned Pigeon and the Horned Iguana, all of which are protected by law. It also embraces diverse flora and has some caves in which pre-Hispanic indigenous people, the Taino, left petroglyphs and petrographs. The caves were spaces in which numerous spiritual-mystical rituals took place, and at least one is closed to the public because of its precious contents. It is with these assets in mind that a project has been proposed to promote ecotourism in the region, specifically based on tours into the park interior, visiting caves, looking at the flora and fauna, learning about the local ecology and exploring the natural coastline.

In common with the other national parks in the Dominican Republic, Del Este is managed by a government agency, the National Parks Directorate (DNP). It oversees all policy and planning, and employs guards to police the entrances and boundaries, ensuring that rules are adhered to and no illegal activity is occurring. Del Este itself was developed after the owner of the land, the Gulf and Western Corporation, agreed to donate it to the country providing it was treated as a protected area. This was in the 1970s and the government also sequestered additional surrounding lands but has not yet compensated all those who lost property. There were a number of smallholdings within the park's designated area, and others regularly used the land for harvesting natural goods including coconuts, honey and animal life (Guerrero and Rose, 1998; Macleod, 2001).

Other organisations have an interest in the park, including USAID, who have financed some of the programmes to help preserve the area, along with The Nature Conservancy organisation (TNC) who have designated it as a 'Park in Peril'. More relevant is the fact that a local NGO has been created to oversee the environmental conservation aspects of the region: Ecoparque. This organisation has responsibility to ensure the park is ecologically healthy; it provides a locally sited unit that has regular access to the area and its population. Ecoparque is running conservation awareness projects in nearby communities, and has started a fisherman's co-operative in Boca de Yuma. However, its funding from the US is due to be reduced, and it has been advised to become self-sufficient on ecotourism. As a result, Ecoparque has begun to assess the park and region for its potential to support ecotours.

To this end Ecoparque employed the services of an ecotourism consultant from the US who, over a period of months involving a handful of fieldtrips, drew up a strategy and feasibility study. Ecoparque also has one local employee, a graduate in environmental studies from a US university, who focused on the ecotourism project in November 1999. He had begun to plan a route through the park, and was becoming familiar with the flora and fauna of the forest. It would be his task to act as a knowledgeable tour guide and/or train others. Unfortunately this employee left for other, better paid work elsewhere and was not immediately replaced.

There were numerous other factors that led to problems in the implementation of the ecotours. Primarily, the ecotourists themselves had to be found. It was suggested that a survey be done of tourists staying in the four nearby hotels (with some 500 beds apiece) within a five-mile radius, to ascertain the likely level of interest. Given the relatively small number of tourists actually anticipated on the tours (100 per week: 2 tours x 10 people x 5 days), it was believed that sufficient people would take the tours. However, other problems arose, as the hotels wished to become involved and operate the tours themselves. This was very much against the desires of Ecoparque who hoped to maintain the ecological sanctity of the park and feared the hotels would disrespect regulations and would not offer a sufficiently high quality of guided service. This situation meant that it would prove difficult advertising for custom within the hotels. Other suggestions included touting the tour to the occasional independent visitors arriving in Bayahibe. Many had shown an interest in the park, and some actually walked there of their own accord (there were no signposts), but discovered that they were refused entry by the guards. In fact, even though the park is advertised as an attraction, the entrance is obscure: about five kilometres miles from Bayahibe village, involving unmarked paths. According to one guard very few independent travellers attempt to enter the park, the only constant visitors being coach parties of students from the country's universities. On the other side of the park, in the east, entry is close to the village of Boca de Yuma, and visitors may walk in unchallenged. By the edge of this village a large cave possesses many well-preserved petroglyphs on its entry walls, and is regularly visited by hotel tour buses, but is not within park boundaries. The Ecoparque employee was not actually aware of the existence of these cave petroglyphs.

In addition to those parties already mentioned involved with the environmental programmes based in the region, there were two US Peace Corps Volunteers (PCVs) working there: one in Bayahibe and the other in Boca de Yuma. This organisation has had an interest in the Dominican Republic for many years, and indigenous organisations often request help from volunteers with specific skills. In this instance Ecoparque together with the Bayahibe Development Foundation had asked for someone with environmental knowledge and experience of ecotourism.

Both the PCVs were young women in their early twenties, recently graduated from college and with an environmental education, as well as experience in Latin American development programmes in one case. However, they both believed that their status was diminished in the eyes of local policy makers because they were women. They said that they were regarded as 'young girls' whose opinions were secondary to the men in the organisation. Further, to one PCV's annoyance, no mention had been made of the ecotourism consultant from the US, whose project had been completed only a few months before her arrival. There were clear problems in communication, trust, respect and co-operation between these organisations and the staff involved.

To add to the staffing and management problems of the ecotourism project, the sense of job insecurity is very high within these organisations, as well as in the tourism sector and public service work. The election of a new national government can signal a complete change in the staffing of public organisations and a consequent change in policy. The future for Ecoparque's initiatives, and their ecotours plan was dependent on the agreement of the DNP with their project. National general elections were held in May 2000, and at the time of fieldwork, some months earlier, the sense of confidence was low. Such transformations impact on every level of staffing, from the upper echelons of management to the park guards.

Not only are the expectations of future planning directly impacted by job insecurity and 'cronyism', but there is a pervasive sense of cynicism throughout the services. For example, many guards are on very low wages and desirous to quit their jobs. They are rarely trained either to interpret the environment or in environmental conservation. There are suggestions that they are open to bribery and corruption, turning a blind eye to hunters/fishermen/farmers for a favour. The same applies to the navy that patrols the waters. It is generally acknowledged that 'money talks', and most of the money is currently involved in the hotel business, where the influence consequently lies.

And so, it has been revealed by Ecoparque that one of the hotels plans to dredge the Caribbean side of the channel for white sand in order to build up and beautify its own inadequate beach. This would accelerate the denudation of marine life in the channel and therefore in the area, on which so much depends. Even more worrying for local conservationists and others, was the growing commercial lobby to sell off parts of the park for hotel development – as well as turn the island of Saona into an exclusive resort. This scenario provides the economic background against which the tiny organisation of Ecoparque operates, with its suggested low-impact ecotours bringing in a minute fraction of the income earned daily by the boat-tours to Saona (some of which goes directly into the DNP coffers). The pressure to create wealth for this relatively poor country is constant, and the success of mass package tourism drives the hotel industry to grab the beauty spots on the coast – to the detriment of local ecology.

Conclusion

It is necessary to examine ecotourism in practice to understand the grass-roots application of this broad-ranging type of tourism in a rural setting. It is of increasing importance in the current global situation, where environmental destruction and global warming is of paramount concern. Studies have revealed that ecotourism is diverse, but most importantly, that the social landscape and cultural context is of absolute relevance to the development of ecotourism. The two cases, spontaneous and planned development, prove to be a useful dyad, comparable with Cohen's notions of the 'organic' and 'induced' (in Wilson, 1993). Both these types of development have positive and negative attributes and illustrate the variety of outcomes, increasing our critical awareness of potential problems. Here it can be seen how ecotourism in different locations, influenced by local factors, is strongly subject to the socio-economic and political environment.

The examination of La Gomera showed that ecotourism is able to develop where tourists have a propensity to respect the natural environment, and have a genuine interest in it – even though their primary motivation is to relax. The 'raw material' is present on which a successful ecotourism project can be built. It might be said that this was a latent asset. Other important factors include the entrepreneurial experience and capital backing, on a small scale, of those individuals who developed the business, and crucially, their flexibility and ability to recognise viable options and markets. This knowledge of the tourists, the region and local political scene is very useful (Smith, 1989). As an example, the maritime centre owner who purchased a fishing boat was a German who had resided in the village for some years and was prepared to wait out the long, bureaucratically frustrating period before being allowed to go ahead with his business venture. He knew the type of tourists well, and furthermore, had an intimate understanding of the local community and maritime resources.

The Dominican Republic example illustrates how an ecotourism project, with the initial support from numerous influential and experienced bodies, can encounter serious problems due to the socio-political and economic environment in which it is based (Olwig, 1980). The organisations USAID, The Nature Conservancy, DNP, The US Peace Corps and Ecoparque were all supporters of the project. But competition 'on the ground' from huge hotels looking after their own financial interests, combined with an orientation among visiting tourists towards the 'sun and sand' all-inclusive hotel vacation, is one factor working against the tours. Others of equal importance are the general working conditions of employees, leading to a high staff turnover, poor quality of education and inexperience. A lack of dedication among those in positions of responsibility also leaves them open to corruption. These factors are combined with poor communications between institutions and personal prejudices, as well as a political climate supporting neo-liberal commercialisation and independence, not immediately oriented towards conservation.

This chapter has shown that the two sites of research are markedly different: they illustrate the diversity of ecotourism initiatives in contrasting settings. Rural tourism, in the form of ecotourism, is playing an increasingly important part in their development, both planned and spontaneous. Field research has illuminated the directions in which rural tourism is heading in terms of its impact on local communities, different types of growth, the character of the tourists and their inclinations. The case studies point to ubiquitous issues and these include the following:

1 the need to conserve the environment and consider the needs of the host community – holistic sustainability – for successful ecotourism development;
2 the existence of local and global pressures for ecotourism to commercialise and become viable;
3 the necessity to understand the social environment (host culture at local and national levels) in which the ecotourism will take place; and
4 the growing importance of rural tourism in the development of peripheral communities around the world.

An awareness of these issues not only allows us to appreciate the importance and complexity of ecotourism in its rural setting, but should lead to more realistic and sensitive developments in the future, beneficial for the natural environment, the local community and the visitors.

References

Brandon, K., Redford, K. and Sanderson, S. (1998), *Parks in Peril: People, Politics and Protected Areas*, Island Press, Washington DC.
Guerrero, K. and Rose, D. (1998), 'Dominican Republic: Del Este National Park', in K. Brandon, K. Redford and S. Sanderson (eds), *Parks in Peril: People, Politics and Protected Areas*, Island Press, Washington DC, pp. 193-216.
Macleod, D.V.L. (1993), 'Change in a Canary Island Fishing Settlement, With Reference to the Influence of Tourism', University of Oxford, unpublished PhD thesis, Oxford.
Macleod, D.V.L. (1997), '"Alternative" Tourists on a Canary Island', in S. Abram, J. Waldren and D.V.L. Macleod (eds), *Tourists and Tourism: Identifying with People and Places*, Berg, Oxford, pp. 129-48.
Macleod, D.V.L. (1998), 'Alternative Tourism: a Comparative Analysis of Meaning and Impact', in W. Theobald (ed.), *Global Tourism: the Next Decade*, Butterworth Heinemann, Oxford, 2nd edn, pp. 150-68.
Macleod, D.V.L. (1999), 'Tourism and the Globalization of a Canary Island', *Journal of the Royal Anthropological Institute*, vol. 5, pp. 443-56.
Macleod, D.V.L. (2001), 'Parks or People? National Parks and the Case of Del Este', *Progress in Development Studies*, vol. 1, pp. 221-35.
Mieczkowski, Z. (1995), *Environmental Issues of Tourism and Recreation*, University Press of America, Boston.
Olwig, K.F. (1980), 'National Parks, Tourism and Local Development: a West Indian Case', *Human Organisation*, vol. 39, pp. 22-31.

Smith, V. (1989), 'Eskimo Tourism: Micro Models and Marginal Men', in V. Smith (ed.), *Hosts and Guests: the Anthropology of Tourism*, University of Pennsylvania Press, Philadelphia, pp. 55-82.

Western, D. (1993), 'Defining Ecotourism', in K. Lindberg and D. Hawkins (eds), *Ecotourism: a Guide for Planners and Managers*, The Ecotourism Society, Vermont, pp. 7-11.

Wilson, D. (1993), 'Time and Tides in the Anthropology of Tourism', in M. Hitchcock, V. King and M. Parnwell (eds), *Tourism in South-East Asia*, Routledge, London, pp. 32-48.

WTO (World Tourism Organisation) (2001), *WTO-UNEP Concept Paper – International Year of Ecotourism 2002*, WTO, Madrid
<http://www.world-tourism.org/sustainable/IYE/WTO-UNEP-Concept- Paper.htm>.

Chapter 15

Relationships Between Rural Tourism and Agrarian Restructuring in a Transitional Economy: The Case of Poland

Lucyna Przezbórska

Introduction

The economic base of rural areas used to be farming, but the problem is that agriculture nowadays needs less labour than it used to and the rural economy has become now much more diverse (Mahé and Ortalo-Magné, 1999). As rural economies are being restructured, there is less emphasis on traditional goods producing sectors such as agriculture, forestry, fishing, and rural manufacturing as ways to provide employment and generating income. That is why part-time farming is increasingly popular, especially among industrial countries and this becomes an important factor to determine on-farm pluriactivity (Ohe and Ciani, 2000). Within farming itself, farmers can increasingly be regarded as 'rural entrepreneurs' who produce a whole range of goods in addition to agricultural commodities and provide a range variety of services (European Commission, 1997). The importance of small-scale entrepreneurship has increased in rural areas and small-scale tourism is intuitively perceived as a suitable form of economic development for rural and agricultural population. Increasingly rural communities of different countries are seeking to utilise recreation, tourism, and tourism related activities to diversify their economies and to replace traditional agriculture related industries which have been obsolete or have left the community. However in many rural areas recreation and tourism have already become the mainstays of the economy (Butler, 1998).

On the other hand, for CEE one of the consequences of economic development into the market economy is a relative decline in the importance of agriculture, including agricultural income decline and increase of unemployment in rural areas. This always has serious economic implications for rural populations. This way the economic environment for farming has forced farmers to consider diversifying into alternative uses for their resources: land, buildings, and others. Development of farm-based tourism provides one possibility for attacking the problem. This way tourism can be seen as an avenue to achieve employment, income generation, and economic stability while providing new uses for older facilities and often providing a focal point for community activity. Tourism is also a way of restructuring and revitalising rural areas and plays an important role when small farms adjust to the decreased prices and increased competition, because through tourism activities agricultural income becomes only one component of the total income (Kaila, 1999). The share of agritourism and rural tourism in the total income of rural households differs between countries, regions, and operations. In some countries, like Austria, rural tourism and agritourism are well developed and there are a lot of farms that benefit even over 50 per cent of income from the activity. On the other hand, Kaila's survey points out that the profitability of pluriactive farm enterprises can be weaker than of traditional farms, especially when they are smaller than the traditional ones. However the off-farm activity is often an important source of income for them. There is also another important point of pluriactivity of farms. Their income diversification is also used to decrease annual income variation (Kaila, 1999; Mahé and Ortalo-Magné, 1999). Agritourism, like no other area of the economy, creates the possibility of enhancing farm income and leads to improvement of living and working conditions of rural population.

In literature there is a great variety and different meaning of terminology and defining tourism in rural areas seems to be complex (Hegarty and Ruddy, 2002). Because of the fact there is a necessity to introduce definitions of agritourism and rural tourism used in the chapter concerning the relationship between rural tourism and agriculture. The term 'agritourism' refers to all tourism and recreation activities connected with a working farm or any agricultural, horticultural, fishery or agribusiness operation. Rural tourism can be defined as a multifaceted activity that takes place in an environment outside heavily urbanised areas (within rural areas), which is different to agritourism.

Agritourism and Rural Tourism in Poland

Many regions of Poland seem to have suitable natural and landscape conditions for agritourism and rural tourism development. The country features nearly all forms of landscapes, including coastal, mountains, lakes, rivers, forests, and uplands that can be attractive as vacation and recreation destinations. Many regions are characterised by high environmental quality, while the number of national parks, landscape parks and nature reserves is increasing (at present there are 1,251 nature reserves, 23 national parks and 120 landscape parks). Tourist services in rural areas have a long tradition going back to the nineteenth century, when various forms of summer recreation expanded in rural areas. After nearly 40 years of interruption, steady development of agritourism and rural tourism has been taking place. Rural tourism draws extensively on regionally-specific conditions and creates jobs for the local population. The Polish Federation of Rural Tourism, 'Guest Farms', established in 1996 brings together more than 45 regional associations and more than 2,200 rural tourism and agritourism enterprises. However, the total number of rural tourism and agritourism enterprises estimated by the Ministry of Agriculture and Rural Development amounts to more than 11,000, including about 5,800 agritourism and 5,500 other rural tourism enterprises. According to the data there are about three agritourism farms per every 1,000 agricultural farms in the whole country. Geographic distribution of the farms differs regionally. Most are located in three provinces: two in the north and north-east (on the coast and the Mazury Lake District) and one in the south (the Carpathian Mountains) (Figures 15.1 and 15.2).

Since 1997 the Federation has been a member of EuroGites – the European Organisation for Rural Tourism. The Federation has introduced detailed guidelines for assigning categories to agritourism lodging facilities. In 2000 there were 685 categorised rural tourism and agritourism enterprises.

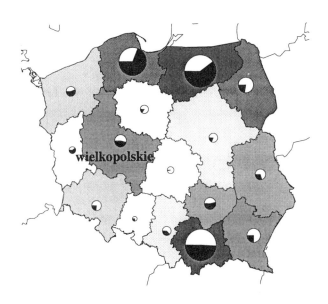

Key: Proportion between agritourism and rural tourism enterprises:

rural tourism   agritourism

Source: Author's own calculations based on data of the Ministry of Agriculture and Rural Development

Figure 15.1   Geographic distribution of agritourism and rural tourism enterprises in Poland, 2000

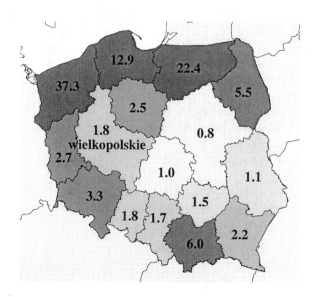

Source: Author's own calculations based on data of the Ministry of Agriculture and Rural
Development

**Figure 15.2   Number of agritourism farms per 1000 of agricultural farms in
Poland by regions, 2000**

The Case of Wielkopolska

Wielkopolska is a large province ('voivodship') in the western part of Poland and
is regarded as an agricultural-industrial region. Development of tourism activities
in rural areas accelerated in the 1990s as a result of introducing a market economy
and restructuring the Polish agricultural sector. Most of the conditions of the region
seem to be favourable for rural tourism development (Table 15.1). Yet tourism
services (including agritourism and rural tourism) are not well developed and the
region is not a traditional tourism destination. In 2000 there were just 460 small-
scale tourism enterprises in the rural areas of the region connected with agricultural
farms and rural households.

 According to different criteria, between 42.5 and 51.3 per cent of Wielkopolska
province's population live in rural areas (Tables 15.1 and 15.2) and about 28 per
cent of the population is connected directly or indirectly with farming (26.5 per
cent of employed population work in agriculture, hunting and forestry – Table
15.1). The average size of an agricultural farm in the region is about 10.1 ha of
agricultural land. Small-scale family farms do not provide opportunities to
accumulate sufficient income to cover their basic needs. The problem is one of the
biggest challenges facing the country as it seeks to modernise and restructure the
agricultural sector without destroying rural communities.

Table 15.1 Characteristics of rural areas of the Wielkopolska province and Poland as a whole (2001)

Specification	Wielkopolska province	Poland
Total area (km^2)	29826	312685
% of Poland	9.5	100.0
Total population ('000)	3366.0	38632.5
% of Poland	8.7	100.0
Rural population ('000)	1429.2	14785.2
% of total population of the region	42.5	38.3
Employed in agriculture, hunting and forestry ('000)	347.2	4289.7
Employed in agriculture in total employed population (%)	26.5	29.2
Registered unemployment rate in rural areas (% annual average in 2001)	16.4	14.8
Number of private farms ('000)	144.0	1881.6
Average farm area (ha of agricultural land)	10.1	7.1
Number of **rural tourism** enterprises (Dec 2000)	212	5471
Number of **agritourism** enterprises (Dec 2000)	249	5789
Number of agritourism enterprises per 1000 farms	1.8	3.1

Source: Central Statistical Office of Poland and author's own calculations

Table 15.2 Rural area definitions applied to the Wielkopolska region and to Poland as a whole (2000)

Criteria of rural areas identification	Specification	Per cent share of rural areas, calculated basing on data at the level of community ('*gmina*')		Average population density (inhabitants per 1 km^2)
		Population	Area	
Administrative recognition of villages and towns (by the Polish Central Statistical Office-GUS)	Poland	38.3	93.4	50.4
	Wielkopolska province	42.5	95.0	50.2
Population density up to 100 inhabitants / 1 km^2 (according to EU methodology)	Poland	32.8	83.0	48.9
	Wielkopolska province	44.0	84.5	58.6
Population density up to 150 inhabitants / 1 km^2 (according to OECD methodology)	Poland	35.0	91.7	47.1
	Wielkopolska province	51.3	91.1	63.4

Source: Ministry of Agriculture and Food Industry and author's own calculations based on data of the Central Statistical Office of Poland, 2001

The rest of this chapter is mainly based on the results of a questionnaire investigation conducted in the Wielkopolska province. It examines the regional diversification of agritourism and rural tourism enterprises, including their connection with agriculture in the past and at present, the influence of tourism activities on farming as well as the income situation. It will show the changes within farming under the influence of agritourism development and the relationship between tourism activities and farming.

Population and Geographic Spread of Agritourism and Rural Tourism Enterprises

Altogether 183 rural tourism and agritourism enterprises of the Wielkopolska region, which started their activity before 2000, were interviewed using a standardised questionnaire survey. The number of such operations is continuing the upward trend, but at present the share of agritourism farms in the region as a proportion of the total number of agricultural farms is still very low, accounting for just 1.8 per thousand (Figure 15.2). The geographical spread of rural tourism and agritourism enterprises across the region is very uneven. Most are situated in the western and north-western areas of the region, where conditions for tourism development in rural areas are most favourable, with a high share of the region's forests and lakes.

Profile of the Enterprises and Entrepreneurs

Most of the enterprises started their tourism activity in the 1990s during which period the economic situation of rural areas dramatically declined. Some 68.8 per cent of the total number of enterprises had provided tourism services for just 2-5 years (Figure 15.3).

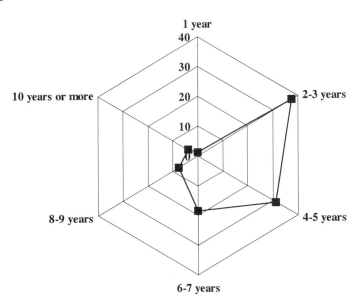

Figure 15.3 Frequency of age groups of rural tourism and agritourism enterprises in Wielkopolska survey (2000)

The total number of rural tourism and agritourism entrepreneurs who took part in the survey was 298. Most of them were married (75.8 per cent of the owners in 61.8 per cent of the surveyed enterprises). Single females were the owners of 20.2 per cent of the enterprises and single males 18.0 per cent. The prevailing majority of the entrepreneurs were between 41 and 50 years old (Figure 15.4). The average age of surveyed population was 45 and the median age was 44. The structure of the owners by gender and age reflects the structure of the whole society of the region: more women than men and more middle-aged people than very young and old. Of course, it can also be noted that both young and old people are not well positioned for undertaking rural tourism or agritourism activities. More than a half of the surveyed population (73.5 per cent) was quite well educated as they had completed at least secondary school.

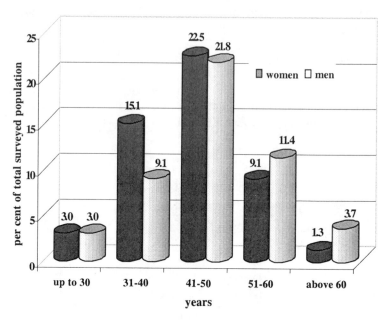

Figure 15.4 Wielkopolska survey: entrepreneurial classification by age and gender

Connections between Agritourism and Farming

From the total number of 183 questioned enterprises 81.4 per cent had agricultural land of over 1 ha and were working farms (Table 15.3). The majority of farms were between 1.1 and 5.9 ha (24.6 per cent) and the average area of agricultural land per active farm was 19.0 ha. That is almost double the average figure for all farms in the region (compare Figure 15.5 and Table 15.1). 18.6 per cent of surveyed enterprises had no agricultural land or had less than 1 ha, while on the other hand the biggest farm was 364 ha.

Table 15.3 Rural tourism and agritourism enterprises and their agricultural characteristics (2000)

Agricultural profile of enterprises	Number of farms	Per cent of total number of surveyed farms
Over 1 ha of total land	167	91.3
Over 1 ha of agricultural land	149	81.4
With fruit and vegetable production	115	62.8
With commodity fruit and vegetable production	71	38.8
With animal production	137	74.9
With commodity animal production	96	52.5
With arable production	151	82.5
With commodity arable production	126	68.9
TOTAL number of surveyed tourism enterprises	183	100.0

Source: Author's own figures based on the questionnaire survey

According to the collected data 62.8 per cent of the total surveyed tourism enterprises had fruit and vegetable production and 74.9 per cent of them had animal production. However not all of them sold their production to the market. Only 68.9 per cent of them had commodity production, including 38.8 per cent with fruit and vegetable commodity production and 52.5 per cent with animal commodity production (Table 15.3). The rest produced agricultural goods for their own needs and for their guests.

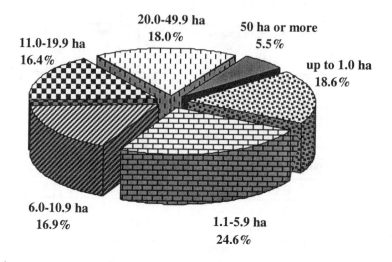

Figure 15.5 Surveyed farms by agricultural land classes

The farms mainly produced cereals, including rye, wheat and triticale (54.6 per cent of farms grew rye, wheat, and triticale). The next position in global and market production had root crops, especially potatoes (produced by 96.6 per cent of all farms with root crops production) and vegetables, including: tomatoes, cucumbers, asparagus, and others. The other plant goods were produced in a small number of farms (e.g. oil crops in ten farms, leguminous plants in ten farms, fruits in eight farms and other plants in six surveyed farms).

Animal production led in 72.1 per cent of all enterprises and three-quarters of them had animal commodity production. They usually produced pork (almost half of the total number of farms with animal production) as well as poultry (40.3 per cent of all farms with animal production). Commodity production was dominated by pork. Some farms also produced beef and milk. Other commodities were produced in a small number of farms and on a small scale.

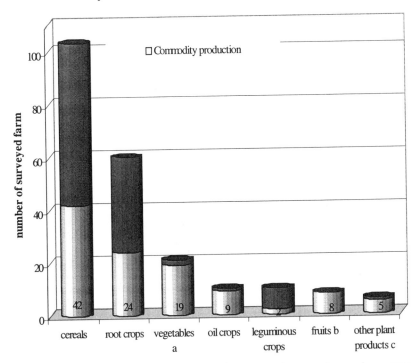

ᵃ including: cabbage, tomatoes, cucumbers, parsley, carrots, beans, onion, and asparagus
ᵇ including: peaches, plums, raspberries, and strawberries
ᶜ including: herbs, mushrooms

Figure 15.6   The surveyed farms by global and commodity crop production

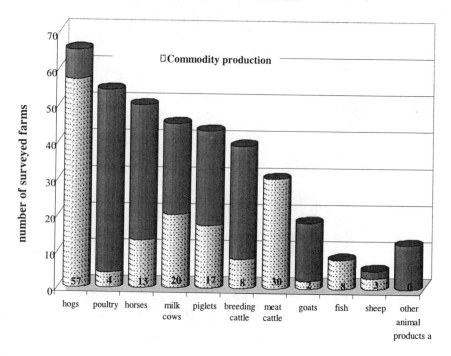

ᵃ including: rabbits, Vietnam pigs, wild boars, and pheasants

Figure 15.7   The surveyed farms by global and commodity animal production

The entrepreneurs were also asked about their connection with agricultural production in previous years, especially from the moment they started tourism activities (Table 15.4). 162 enterprises (88.5 per cent of the total number) had been agricultural farms in the past and 158 were farms when they started tourism activity. In 1999, the year before the surveyed year, 155 of the enterprises produced agricultural goods at least for themselves (84.5 per cent of the total number of operations) and in 2000 149 were still farms. The reasons for reducing or leaving farming varied. Most of the entrepreneurs had derived low or no income from farming, some quit farming because of deteriorating health or retirement, and only the owners of one enterprise decided to reduce farming because it was too labour-consuming combined with tourism services.

The entrepreneurs also indicated the influence that tourism activities had on the changes in their farms or agricultural production (Table 15.4). More than a half of the entrepreneurs claimed that tourism activities influenced farming and 42.6 per cent of tourism enterprises' owners had introduced new agricultural production adjusted to the needs of agritourism development. The most frequently mentioned were horses (20 farms), poultry or goats (24 farms), growing vegetables and fruits (usually emphasising organic production), and six farms had started fish production. The aim of introducing new or changed directions of production was to ensure contact with live animals for children as well as to provide fresh food products that could be used to prepare meals for guests. Several farms have kept animals such as rabbits, sheep and different species of poultry (including peacocks, turkeys, pheasants) especially for tourists with children. The introduction of tourism activities was also an incentive to re-arrange and tidy up their backyards and the surroundings of their houses.

Table 15.4   Changes within farming introduced because of agritourism activity

Changes within farming	No of answers/ farms	Per cent of surveyed operations
Introduction of new agricultural production, including:	78	42.6
Vegetables	13	7.1
Fruits, orchard	8	4.4
Poultry	13	7.1
Special races of poultry, including: hens, turkeys, pheasants, peacocks	2	1.1
Horses	20	10.9
Goats	9	4.9
Cows	1	0.5
Pigs	2	1.1
Rabbits	1	0.5
Sheep	2	1.1
Fish (fish ponds)	6	3.3
Others	1	0.5
Limitation of agricultural production	4	2.2
Resignation of agricultural production	4	2.2
Resignation of animal production	5	2.7
Arrangement of the area of a farm	2	1.1
Adjustment of agricultural production to agritourism activities	2	1.1
The whole farm adapted to agritourism activities	4	2.2
Land purchase (for recreation)	1	0.5
TOTAL:	100	54.6

Source: Author's own data based on the questionnaire survey

The entrepreneurs tried to outline their plans for future in relation to farming. The owners of 140 enterprises were sure they would still produce agricultural goods in the future, 36 entrepreneurs said categorically no for farming, and seven were undecided.

Sources of Income

Most of the surveyed households connected with rural tourism and agritourism enterprises had usually more than one source of income (Figure 15.8). First of all most of them benefited from farming and tourism services, but also from other off-farm small-scale enterprises, such as trade and services for agriculture, or from non-earned sources of income such as retirement and unemployment benefits. More than 80 per cent of the interviewed households had two or three sources of income; there were some with four or five sources and one with six different sources of income.

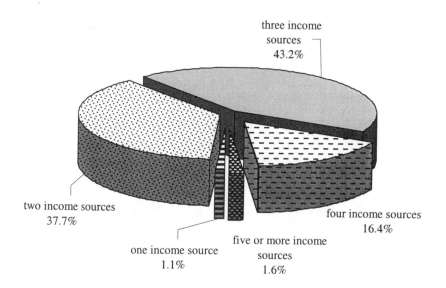

three income
sources
43.2%

two income sources
37.7%

four income sources
16.4%

one income source
1.1%

five or more income
sources
1.6%

Figure 15.8   Surveyed rural tourism and agritourism enterprises by the number of income sources

In terms of enterprises' self-defined main additional sources of household income (Figure 15.9), farming was still the most often mentioned main source. 81.4 per cent of the total surveyed enterprises were still connected with farming and for 45.4 per cent of all respondents (55.7 per cent of the agritourism ones) farming was the most important and the main source of income. Contracted off-farm work for one or more household members was the main source of income for 22.4 per cent of households. A relatively significant proportion of the interviewed population (16.9 per cent) treated their rural tourism or agritourism activity as a main source of income for the whole family. A similar share was indicated for non-earned sources of income, including retirement or unemployment benefits, and self-employment.

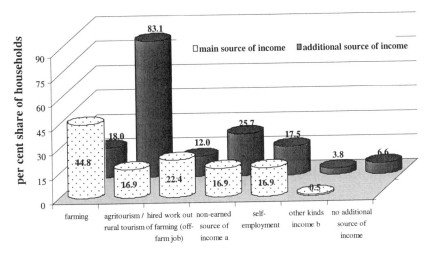

^a retirement, social benefits, pre-retirement benefit, unemployment benefit
^b seasonal work in Germany

Figure 15.9   Main and additional sources of income of the surveyed farms

The entrepreneurs also indicated their additional sources of income (Figure 15.9). It can be seen that for 83.1 per cent of surveyed households agritourism or rural tourism was the most important additional source of income. Over one quarter of interviewed households derived their additional income from retirement or unemployment benefits and this source of income was very important for them because they received it regularly every month. In 96 households at least one person received such benefits (in 35 operations two people and in one household even three people). For 17.5 per cent of the households farming was an additional source of income.

Next is an evaluation of the surveyed population's estimates of the per cent share of agritourism or rural tourism income in the total income of their households (Figure 15.10). In general the share of tourism activity income was at an average level. Most often it was between 6 and 10 per cent (21.3 per cent of households) and between 11 and 20 per cent (17.5 per cent of households). However for 14 households tourism activity income did not exceed one per cent of the total income and for 29 per cent of them it was lower or equal to 5 per cent of the total income. On the other hand seven agritourism or rural tourism entrepreneurs estimated the share of tourism income in their total income at more than 75 per cent.

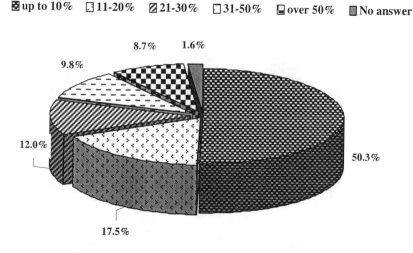

Figure 15.10 Share of agritourism/rural tourism income in the total income of the interviewed households

Types of Rural Tourism Enterprises

Analysis of collected data provided a three-fold classification of the agritourism and rural tourism enterprises in relation to their agricultural production function:

1 type I: enterprises directed at tourism activity as the main source of their income, with quite large agricultural areas and agricultural commodity production (most often animal production), usually connected with equine recreation;

2 type II: typical small-scale family farms which make a living by agricultural production, and agritourism is a source of additional income and use of some additional free accommodation; and

3 type III: very small agricultural farms or rural households with different sources of income, for which tourism activity is often the main or only source of income.

Because of a quite weak interest in rural tourism and agritourism held by potential Polish tourists (few guests, short tourism season), the existing accommodation base and attractions seem to be sufficient to satisfy current demand for this kind of service. In the future, however, taking into account tourism trends in most West European countries and the tendency of Polish consumers to follow them, we can expect further development of agritourism and rural tourism attractions and growth in the number of rural tourism and agritourism service providers.

Conclusions and Observations

The data and analysis presented in this chapter lead to the following concluding observations:

- tourism enterprises in rural areas of Poland are still very dependent on farming and farming is still the main source of income for most of them. However, some of the entrepreneurs have tried to change or adjust farming to tourism activities, but the changes are rather insignificant. Agricultural farms require more time to change their profiles;
- most of the surveyed enterprises were pluri-active and drew on income from several different sources. The entrepreneurs are very active in seeking new sources of income for their households (only two of them had only one source of income). Rural tourism or agritourism activity was a good experience for the surveyed families as it permitted them to learn how to manage their own businesses;
- the share of agritourism or rural tourism income in the total income of surveyed families is not very high, although given the short time devoted to providing tourism services by most of the enterprises it can be recognised as acceptable. Most of the surveyed population was satisfied with their tourism income. The small share of agritourism/rural tourism income in the total income of the households was mainly a result of quite small numbers of guests, especially during the first years of activity. Polish people are not used to visiting agritourism or rural tourism farms during their holidays or vacations and when they go to the countryside many urban visitors stay with relatives; and
- restructuring rural areas and agricultural farms through rural tourism and agritourism development seems to be easier than through the other means and other activities. However, it is a long process requiring much financial investment not only within farms themselves but also within the wider rural sector.

References

Butler, R. (1998), 'Rural recreation, and tourism', in B. Ilbery (ed.), *The Geography of Rural Change*, Addison Wesley Longman, Harlow, pp. 211-32.

European Commission (1997), *Rural Developments*, CAP 2000 Working Document, Directorate General for Agriculture DG VI, Brussels.

Hegarty, C. and Ruddy, J. (2002), 'The Role of Rural Tourism Entrepreneurship in Regional Development', Re-inventing a Tourism Destination, TOURISM 50th Anniversary Conference Proceedings.

Kaila, M.M. (1999), 'Economic results of pluriactive farm enterprises', VII European Association of Agricultural Economists Congress, Warsaw.

Mahé, L.P. and Ortalo-Magné, F. (1999), 'Five proposals for a European model of the countryside', *CAP and the countryside, Proposals for the food production and the rural development, Economic Policy, A European Forum*, No 28/1999, Centre for Economic Policy Research, Centre for Economic Studies, Maison des Sciences de l'Homme, Blackwell, London.

Ministry of Agriculture and Food Industry (1999), *Spójna polityka strukturalna rozwoju obszarów wiejskich i rolnictwa*, Ministry of Agriculture and Food Industry, Warsaw.

Ohe, Y. and Ciani, A. (2000), 'On-Farm Tourism Activity and Attitudes of the Operators a Hiroshima – Umbria Comparative Case Study', The Technical Bulletin of Faculty of Horticulture, Chiba University, vol. 54, pp. 73-80.

PART 5
CONCLUSION

Chapter 16

New Directions in Rural Tourism: Local Impacts and Global Trends

Lesley Roberts, Morag Mitchell and Derek Hall

The eclecticism evident in the preceding chapters reflects the diversity of rural tourism production and of the processes that influence its development in both local and global contexts. Indeed, the diversity of rural tourism, together with its fragmentation and the intangibility of its characteristics represent both its essence and its challenges, and go some way to explaining its lack of clear presence in both the international rural development and policy arenas.

The conference from which most of these chapters evolved had three main aims. These were to:

1   identify key contemporary research issues;
2   highlight areas of policy and practice that might be informed by research; and
3   explore likely future dimensions of rural tourism development and practice.

By way of conclusion therefore, this final chapter aims to reflect a convergence of these conference aims and the research presented by the authors in this volume, highlighting key issues and relating them to the policy development concerns of rural tourism stakeholders – rural communities, businesses and development agencies.

The Need for Rural Development Policy

Rural development occupies an increasingly important place in policy debates across Europe, and indeed within the global context, although there is as yet little evidence of integrated policy. In 1996, the final declaration of the Cork conference on rural development in Europe defined policy objectives to be pursued as:

* the reversal of rural out-migration;
* combating rural poverty;
* the stimulation of rural employment and equality of opportunity; and
* responding to growing requests for quality, health, safety, personal development and leisure, and well-being,

whilst recognising the need to conserve and enhance rural qualities and environments.

The conclusions of the Cork Declaration have been pursued largely through Agenda 21, and the European Commission has made the development of sustainable rural development policy a priority. Through the LEADER programme for example, currently in its third phase, support has been given to rural development projects designed and managed by local partnerships. Such an explicit territorial approach to policy delivery has contrasted with the more usual sectoral approach which has characterised policies attached to agriculture and, more latterly environment. The search for effective means of integrating the multiple and dynamic needs of contemporary rural areas has thus achieved more prominence.

Globally, the roles of rural tourism and recreation industries within integrated rural development are well understood in principle. Considerable attention has been given to the support and enhancement of rural tourism initiatives (Mormont, 1987; Bethemont, 1994; Hjalager, 1996; Priestley *et al.*, 1996; Edgell, 2002) in the wider context of rural development. Tourism is seen as being of considerable economic and social benefit to rural areas through the income and infrastructure developments it may bring to marginal and less economically developed regions. Rural tourism and recreation industries support economic and social restructuring. They play a vital part in the dynamics of rural populations, and are instrumental in the creation and development of new communities, both at a local level and more widely by way of communities of interest (ETC/CA, 2001: 5; Roberts and Hall, 2001: 6). Crucially, however, rural tourism industries' potential is only likely to be maximised by inclusion in integrated development plans, their multi-sectoral nature contributing broadly across both regions and sectors.

However, while roles may be well understood in principle, the ways in which they might be achieved in practice have been, and remain, uncertain. The restructuring of agriculture, the emergence of rural development policies, and changes in rural consumption for leisure purposes all contribute to significant rural change, and the relative lack of recognition of these important changes as they interact with rural tourism and recreation reflect an indifference to the industries that almost borders on neglect. These are critical relationships that remain largely unquantified and are at best implicit in rural policy documents. Often, they are ignored.

Globally, however, a concept of 'countryside capital' is emerging that may be identified as the foundation of much successful rural economic activity. The fabric of the countryside with its distinctive landscapes, biodiversity, historic features and the built environment of villages and market towns is both the product and the source of such capital, providing us with yet another term with which to grapple for clear definition and understanding. In this context, 'capital' may be defined as capacity that provides resources for development, and it may be identified in a number of forms: economic, natural, human, social and cultural (Donald Macleod's account, in Chapter 14, points to the importance of different capital forms in the development of ecotourism). It is generally understood that strength across the range is required for sustained economic growth and development. Again, however, although we are comfortable with the concept(s) in principle, the means by which they may be translated into useful policy or management direction remain unclear. Further research is required into the relevant importance of each form of capital, the means by which each might be achieved, the inter- and co-dependencies that emerge, and their complex relationships with development (which comes first – development or capital?).

For rural tourism, countryside capital operates at two levels – as a backdrop within which 'pure' forms of rural tourism (Lane, 1994, 1999) take place, and as a resource for accommodating the range of contemporary tourism and recreational activities that often have little to do with either rural characteristics or values (Butler, 1998: 215). It seems to be widely accepted that 'new' forms of tourism and recreation in rural areas are changing the nature of the industries that collectively make up what are referred to as 'rural tourism' (Butler and Hall, 1998: 249; Swarbrooke, 1999: 169; Roberts and Hall, 2001: 224). Increasing visitor numbers and the changing nature of consumption combine to effect significant modifications in terms of both tourism's scope and scale. We need, therefore, to find ways to identify the mutual values shared by rural tourism and countryside capital, reflecting the bi-directional nature of development processes in the interests of effective policy formulation and development.

Review and Conclusions

What we mean by 'rural tourism' therefore, becomes more than academic debate, as emphasised in the first of the book's five themes, *Context*, in Chapter 1. Whilst recognised at the level of the provider, the significance of various (new) forms of participation may only now be emerging at policy level. Mass tourism in Europe's countryside, like mass tourism on its coasts, may be widely considered to represent a market for private goods whereby supply and demand can achieve equilibrium without the occurrence of market failures. As such, it would warrant no enabling policy intervention – merely a framework for its control. Globally however, the visitor economy represents the sole income generator for many rural areas, and failure to recognise its importance, potential and impacts may be critical for rural development generally. What we mean by 'rural development' is also at question. The OECD describes it as 'a dynamic concept, encompassing multiple objectives such as equalisation of incomes of rural and urban populations, equal access to social services, creation of equal employment opportunities and protection of rural amenities' (OECD, 1998: 81). Rural tourism and recreation industries have the potential both to build on and to undermine such processes if insufficiently understood and inappropriately managed.

The dynamic relationships between tourism and rural development are explored within the second of the book's key themes, *Conceptualisation,* although reflecting the authors' holistic approaches, such relationships are evident throughout the volume. Although agricultural production is still heavily centrally controlled, there is recognition of an increasing plurality and the concept of govern*ance,* rather than govern*ment* in the contemporary politics of much of rural Europe (Goodwin, 1998; Marsden and Murdoch, 1998). In Chapter 3, Richard Sharpley clearly illustrates, through his analysis of the outbreak of foot and mouth disease in the UK, the fragility of tourism within the rural sector. His chapter emphasises the difficulties of following a governance model when central government policies, both direct and indirect, maintain power and control. It shows how current problems being encountered by rural tourism and recreation industries across the UK are the result of extrinsic forces rather than intrinsic shortcomings, reflecting the fragile position of rural areas generally within the wider geography of development. His account reveals the limited extent to which rural activities other than agriculture are formally recognised by those managing and administering the rural sector. The exposure of rural tourism industries to external influences without policy support seriously undermines their ability to survive let alone to contribute positively to future development. Anders Sørensen and Per-Åke Nilsson, in Chapter 4, draw attention to the unhelpful distinction between tourism and recreation, a distinction critical to governments' perceptions of the ways in which each can support and be supported by related agrarian activities and by the public sector in general – a point to which this chapter will return. Overall, therefore, policy responses are generally seen as critical to rural tourism's development although there is little agreement on what these responses might be. Despite the rhetoric, however, in practical terms, rural policy development with regard to tourism remains embryonic.

Through the theme of *Experience,* this volume explores the practical dimensions of developments, which may often continue apace despite the lack of policy direction. In Chapter 5, Lesley Roberts and Fiona Simpson analyse the challenges posed by the hedonism inherent in tourism and recreation contexts to attempts to encourage visitors to behave in the interests of the environment, and it is suggested that, through personal contact, land managers may have a role to play in the required education and interpretation processes. Hazel Tucker, in Chapter 6, focuses on the importance of 'exchange' in management of the visitor experience. Mette Ravn Midtgard, in Chapter 8, focuses on perceptions of authenticity within a context of peripherality. In Chapter 7, Derek Hall and colleagues point to the importance of recognition of visitor-animal interaction for business purposes. Such findings illustrate that successful tourism enterprises have moved beyond the concept of service provision to embrace the experience economy (Pine and Gilmore, 1998) with management of the visitor experience as a focus. Further research into the motives of rural visitors and the values they place on the countryside as a recreational resource will inform the process of experience provision. Our understanding of experiences sought is based largely on assumptions that, given mounting evidence of changes in activities and visiting patterns, are little more than supposition. Changing behaviour patterns, lifestyles, demographic shifts, and increasing consumer choice are some of the supposed causes of a fragmentation that have rendered consumption patterns unpredictable (Thomas, 1996). Moreover, it is increasingly recognised that consumption is less about the product/service attributes that confer tangible benefits upon the buyer than the symbolic nature of consumption that defines images and demarcates social relationships (Featherstone, 1991: 16). Indeed, as explored by Glen Croy and Reid Walker in Chapter 9, this can apply to film-induced tourism and other forms of tourism drawn to rural areas through image projection and encapsulation by contemporary popular culture.

Rural tourism enterprises will be better supported by accurate information that allows them to focus on identification and delivery of visitor experience rather than on the sale of services to target markets according to visitor age, origin or (supposed) motivation.

The book's penultimate theme of *Strategy and management* serves to explore some of the practical contexts within which rural tourism must operate. How might government and non-government organisations approach the shift in emphasis from government to governance, for example? In Chapter 10, Hans Embacher's account of the work of the Austrian Farm Holidays Association reveals some of the tensions of working within a federal-provincial network where provincial member organisations exhibit different stages of development and are thus unable to operate on an equal footing. How might we operationalise the conceptual shift from agricultural to rural policy? One starting point for this has been advanced by the OECD (OECD, 2001) which has approached studies of rural multifunctionality through a positive concept that analyses specific characteristics of agricultural production processes and outputs, and recognises the existence of multiple commodity and non-commodity outputs jointly produced by agriculture – for example landscape resulting from livestock-grazing. Critically, recognition is given to the fact that the markets for some of these outputs are likely to function poorly resulting in externalities and market failure. For example, rural areas themselves may satisfy 'option demand' (Johnson and Thomas, 1992: 3) where the existence of the attraction is given added value because it provides potential visitors with the choice to visit even though the choice may never be exercised. Justification for public intervention rests on the belief that the capacity of rural areas to adapt adequately to changing circumstances is slowed by a range of market failures. The economic performance of rural areas often falls below national averages because of economic structure and/or geographic peripherality. The perception that rural problems can be attributed, not necessarily to the intrinsic features of rural areas themselves, but to the ways in which complex sets of national and global forces influence their development (OECD, 1998: 81) supports the existence of market failure and argument for government support.

It is possible to identify a number of market failures of rural businesses operating in the fields of rural tourism, with implications for practitioners and entrepreneurs as well as policy-makers:

- there is no costing of negative externalities accruing from rural tourism businesses, and prices do not reflect either the product's value to the consumer or its effects on the environment. For example, tourists do not normally pay for landscape or heritage feature maintenance;

- diversity, dispersion and fragmentation make it difficult to define sectors, identify their characteristics and needs, and develop co-ordinating structures to support development;
- the industries are characterised by a predominance of family-run businesses with resource inefficiencies and inadequate returns on capital;
- there is often a lack of necessary business and marketing skills to operate in a competitive market economy, and inadequate recognition of the need to work co-operatively and develop a critical mass of facilities and amenities (Hjalager, 1996; McKercher and Robbins, 1998; Clarke, 1999; Roberts and Hall, 2001: 205). See Patricija Verbole's account of training and education needs in Chapter 13.

Due to their development potential, and because of the negative externalities with which they have to contend, it seems clear that tourism and recreation industries in the rural context are valid targets for governments' interest and support. Whilst the forms these will take are likely to vary between countries and regions, information needs are likely to have a shared basis, and include:

- exploration of the links between agricultural, environmental and tourism production, and identification of complementarities, conflicts and co-dependencies;
- establishment of the diversification choices such links create for farming and other rural communities. Are there alternatives to rural tourism that better fit the development needs of the region? In Chapter 15, Lucyna Przezbórska highlights the popularity of tourism as a means of diversification, based on populist assumptions about its superiority over other forms;
- analysis of existing and potential forms of countryside capital, and articulation of their importance to development processes;
- identification of the implications of findings for quality of environment, produce and visitor experience. In Chapter 12, Ray Youell points to the importance of quality provision not just for product development but for wider issues of integration;
- improvements in understanding visitor needs in the interests of experience creation and provision;
- recognition of a 'visitor economy' to embrace leisure, recreation and tourism in order to achieve synergies in management, thus creating balance between demand and supply.

This last point raises the idea of 're-labelling' – with the effect of re-orientating – tourism. This is a potentially a critical issue that broadens the development focus of the industry to a more inclusive concept, further embedding it within the policy domain. 'Recreation', a public good, is often contrasted with 'tourism' which, in terms of direct provision and consumption, is still largely a private good. 'Leisure' would appear to be an even more value-laden term, richly influenced by cultural context, but outside the scope of this debate. Although the impacts of each may vary in intensity (tourist daily spend per head, for example, is usually higher than day visitors'), differences in their nature may be negligible. Data-collection systems that accurately distinguish tourist spend are rare in any case, and negative impacts such as congestion and erosion do not respect the semantic divide.

The 'rural visiting' of the future, therefore, is likely to be understood in terms of the experience it provides for rural economies and their residents, both permanent and temporary. Sustainability depends on a convergence of mutual understanding (social capital). It requires management of an environment that embraces people as well as landscapes and reflects the needs of both. Whilst government agencies will play vital roles in development processes as influencers, enablers and leaders, implementation of a visitor economy should largely depend on 'others' in the public, private and voluntary sectors (ETC/CA, 2001: 18). The importance of networks, collaboration and voluntary participation is therefore not to be under-estimated – see Alenka Verbole's account of partnership building in Slovenia in Chapter 11. At the stage of implementation, issues relating to quality (Ray Youell, Chapter 12) and experience (Hazel Tucker, Chapter 6; Derek Hall *et al.*, Chapter 7; Mette Ravn Midtgard, Chapter 8; and Glen Croy and Reid Walker, Chapter 9) become the responsibilities of a range of stakeholders with local governments or government agencies taking nothing more than a co-ordinating role. Research to enhance understanding of social and cultural capital is essential to the success of such development.

Demand for recreational use of the countryside (whether for 'tourism', 'recreation' or 'leisure') has increased and is likely to increase still further. Despite inconsistent and incomplete data, an emerging pattern internationally shows that visitors are already the largest contributors to many rural economies. Within visitor management programmes, however, insufficient weight is given to conservation and enhancement of the rural environment (Osborn and Crake, 2001) and too much visitor activity is focused on too limited a number of 'honey-pot' destinations. To improve this situation in the interests of all parties, we need to know more about our rural visitors, about their motives and needs, and about the ways in which these might be met by existing and potential rural tourism industries. Indeed, we need to be able to conceptualise in a shared manner about the nature of rural visiting generally – how it is manifested, what it means to both suppliers and consumers, and how much (and what kinds of) change stakeholders are prepared to accept in pursuit of its development.

This brief concluding chapter has raised a number of key issues surrounding the justification for, and nature of, government involvement in rural tourism and recreation as policy foci shift from agricultural or commodity production to rural non-commodity production (and consumption). Clearly, government involvement is required and the question of its nature is critical   The extent to which tourism and recreation industries can contribute to the changes required of integrated rural policy depends, therefore, on the nature of research (see Chapter 2) and the extent to which its findings can inform policy development and support good practice. This echoes the final dictum of the *Auchincruive Declaration* (Hall, 2000). Rural tourism researchers need to communicate their work to inform the inherently complex and often contentious work of policy making and strategy development. For their part, practitioners and policy makers must look to the research community for assistance with the difficult tasks of policy formulation, strategy and business development. The synergies that accrue will serve to maximise the returns from rural visitor economies in the interests of viable rural worlds.

References

Bethemont, J. (ed.) (1994), *L'Avenir des Paysages Ruraux Européens*, Laboratoire de Géographie Rhodanienne, Lyon.

Butler, R. (1998), 'Rural Recreation and Tourism', in B. Ilbery (ed.) *The Geography of Rural Change*, Addison Wesley Longman, Harlow, pp. 211-32.

Butler, R. and Hall, C. (1998), 'Conclusion: The Sustainability of Tourism and Recreation in Rural Areas', in R. Butler, C.M. Hall and J. Jenkins (eds), *Tourism and Recreation in Rural Areas*, John Wiley and Sons, Chichester, pp. 249-58.

Clarke, J. (1999), 'Marketing Structures for Farm Tourism: Beyond The Individual Provider of Rural Tourism', *Journal of Sustainable Tourism*, vol. 7, pp. 26-47.

Edgell, D.L. (2002), *Best Practice Guidebook for International Tourism Development for Rural Communities*, Brigham Young University, Utah.

ETC/CA (English Tourism Council with the Countryside Agency) (2001), *Working for the Countryside. A Strategy for Rural Tourism in England 2001-2005*, English Tourism Council, London.

Featherstone, M. (1991), *Consumer Culture and Postmodernism*, Sage, London.

Goodwin, M. (1998), 'The Governance of Rural Areas: Some Emerging Research Issues and Agendas, *Journal of Rural Studies*, vol. 14, pp. 5-12.

Hall, D. (2000), 'Rural Tourism Management: Sustainable Options Conference', *International Journal of Tourism Research*, vol. 2, pp. 295-9.

Hjalager, A-M. (1996), 'Agricultural Diversification into Tourism', *Tourism Management*, vol. 17, pp. 103-111.

Johnson, P. and Thomas, B. (1992), 'Tourism Research and Policy', in P. Johnson and B. Thomas (eds), *Perspectives on Tourism Policy*, Mansell Publishing, London, pp. 1-13.

Lane, B. (1994), 'What is Rural Tourism?', *Journal of Sustainable Tourism*, vol. 2, pp. 7-22.

Lane, B. (1999), 'What is Rural Tourism? Its Role in Sustainable Rural Development', Nordisk Bygdeturism Nätverk Conference, Kongsvinger, Norway.

Marsden, T. and Murdoch, J. (1998), 'The Shifting Nature of Rural Governance and Community Participation', *Journal of Rural Studies*, vol. 14, pp. 1-4.

McKercher, B. and Robbins, B. (1998), 'Business Development Issues Affecting Nature-based Tourism Operators in Australia', *Journal of Sustainable Tourism*, vol. 6, pp. 173-188.

Mormont, M. (1987), 'Tourism and Rural Change', in M. Bouquet and M. Winter (eds) *Who From Their Labours Rest? Conflict and Practice in Rural Tourism*, Avebury, Aldershot, pp. 35-44.

OECD (1998), *Agricultural Policy Reform and the Rural Economy in OECD Countries*, OECD, Paris.

OECD (2001), *Multifunctionality: Towards an Analytical Framework*, OECD, Paris.

Osborn, D. and Crake, P. (2001), 'What is the Countryside For? A Radical Review of Land Use in the UK', *What is the Countryside For? Paper 1*, The Royal Society of Arts, Press

Release <http://www.thersa.org/projects/project_closeup.asp/[what-is-the-countryside-for[1].pdf]>.

Pine, B. and Gilmore, J. (1998), 'Welcome to the Experience Economy', *Harvard Business Review*, July-August, pp. 97-105.

Priestley, G.K., Edwards, J.A. and Coccosis, H. (eds) (1996), *Sustainable Tourism? European Experiences*, CAB International, Wallingford.

Roberts, L. and Hall, D. (2001), *Rural Tourism and Recreation: Principles to Practice.* CAB International, Wallingford.

Swarbrooke, J. (1999), *Sustainable Tourism Management*, CAB International, Wallingford.

Thomas, M.J. (1996), 'Consumer Market Research: Does it Still Have Validity? Some Postmodern Thoughts', *Marketing Intelligence & Planning*, vol. 15, pp. 54-9.

Index

accommodation 47, 57-61, 81-9, 138-50, 172, 174, 177, 178, 179, 194, 196, 198, 200, 201
Agenda 21, 53, 226
agriculture 6-7, 49-51, 61-2, 76, 138, 146, 169, 177, 193, 205-21, 229-31
agritourism/farm tourism 63, 138-50, 170, 173, 183-4, 187, 205-21, 229
angling *see fishing*
animals/wildlife 5, 11, 42, 45-6, 75, 82, 90-95, 97-101, 121, 125, 194, 196, 198, 214, 216-17
Auchincruive 14, 190, 231
Australia 35, 75, 117-19
Austria 14, 63, 138-50
authenticity 4, 10, 57, 91, 175, 228

behaviour 19, 24, 28, 60-61, 68-79, 80, 83, 86, 88, 91-2, 94-100
blind optimism 11
book towns 116
branding 12, 144-5, 150
British Tourism Authority 123
brochures 106
buzz 83

camping 60-61
Canada 14, 117, 119, 126
Canary Islands 194-203
canoeists 11
Central and Eastern Europe 7, 100
China 117, 126
codes 70-72, 75, 77, 84, 96, 98, 100-101
Common Agricultural Policy 7, 51, 67

community 7, 11-14, 19, 20, 22, 25-9, 33-6, 40, 42, 51-2, 63, 82, 102, 106, 112, 152-167, 172, 174, 176, 178-181, 191, 202-203, 205, 225-6, 231
conservation 5, 12, 42-3, 51, 68, 74-5, 77, 231
Cork Declaration 225-6
Countryside Agency (England and Wales) 38, 42, 52, 70, 78
cronyism 201
culture 6-7, 10, 24, 42, 49-52, 56, 61-3, 81-2, 88, 115-28, 172, 191, 229
Cumbria 47
cycling 60, 194, 197

decision-making 138-50, 152-67
de-industrialisation 115
Denmark 54-6, 58-63, 102
dependency 26, 43
Dominican Republic 194-203

ecotourism 21, 194-203, 226
education and training 12, 145, 179, 183-92, 230
edutainment 90-91
employment 5-6, 8-9, 13, 22-5, 30, 43, 49-50, 54, 90-93, 99, 176, 205-21
England 10, 15, 45, 77, 89, 117, 122-3
English Tourism Council 46, 48, 52, 123
European Commission 169, 226
European Union 4, 14, 51, 197
exchange theory 22, 80

experience 19, 26, 29, 33, 36-7, 40, 42, 47, 53, 57, 60, 69, 73, 76, 80-81, 85-100, 102-112

farm tourism *see agritourism*
film 115-28
fishing/angling 9, 58, 82, 95, 102-112, 199
foot and mouth disease 39, 45, 51-2, 115, 227
forestry 43, 196, 198, 209
France 116, 126

gastronomy 6, 174, 178-9
gate-keepers 164, 166
General Agreement on Tariffs and Trade 46
gentrification 8, 42
Germany 53, 59, 64
globalisation 9, 184, 115, 127, 225-32

hiking/walking 9, 45, 59, 70, 92, 183-4, 194
horse riding/pony trekking 60, 183
hosts 21-2, 24, 26, 30-31, 33, 37, 77, 80-89, 183-92
hunting 60, 95, 161, 209

idyll 8, 10, 12-15, 42, 44, 91, 102-112
impacts 5-6, 13, 19, 21-7, 30, 33-7, 45-8, 62, 72, 74-5, 78, 90, 92, 96, 98, 100, 127, 225-32
income leakage 5
India 100, 119, 125-6
information technology 145-7, 173
integrated quality management 169-81
interpretation 12, 27-8, 72-9, 90-93, 95-101
INTERREG 143
Ireland 15, 101
Italy 121

Japan 81, 126

Lake District (English) 47, 53
landscape 4-6, 8-9, 12, 14, 15, 55, 57-63, 91, 102-112, 229, 231
LEADER 14, 41, 226
Legoland 59
literary tourism 116-17

Malaysia 120
marine tourism 183, 194, 196-9
market failure 229
marketing 11-13, 35, 39, 48, 138-50, 170, 178
migration 4-5, 7, 9, 225
mountain biking 183, 197
muffin 84-5
multiplier 5

national parks 10, 45, 194-203, 206
nature tourism 70, 194-203
networking 152-67
Netherlands 8, 14-15
New Zealand 32, 36, 80-83, 85-6, 89, 97, 101, 115-28
non-governmental organisations 152-67, 195, 199
Northumbria 47, 53
Norway 102-112

one-night stand 86
Organisation for Economic Co-operation and Development 227, 229
outdoor equipment 4

paintballers 11
partnerships 13, 41, 53, 170, 175, 178
peripherality 13, 35, 102-112, 229
petroglyphs and petrographs 199
phenomenonology 34
photographs 106
place marketing 11-12
Poland 205-221
policy 3-4, 7, 11-13, 23, 39, 40-41, 43, 45-6, 48-53, 62-4, 67, 99, 170

pony trekking *see horse riding*
postmodernity 63
post-productivist countryside xi
poverty 225
pubs 47, 154, 156

quality of life 6, 172

residents' perceptions 23, 25-6, 30,
    33-5
resource management 62-3, 183-92,
    194-203, 205-32
restructuring 3, 5-7, 9, 13-15, 35, 52,
    63, 71, 205-21
Romania 117
Rousseau 71
Rwanda 121

Sartre 108
Scotland 14, 19-20, 26-7, 29-30, 36-
    7, 67-70, 72-3, 79, 116, 120, 190
seasonality 21
sense of place 11, 74-6, 96-7
sensual experiences 110
Singapore 126
Slovenia 152-67, 231
social construction 5, 10, 13
Spain 121, 194-203

spirit of place 12, 14
Star Trek festival 122
strategy 38, 46, 52, 138-150
sustainability 6, 12, 14, 19, 39-40,
    50-53, 86, 92, 128, 174-5, 194-
    203, 231
sweeping generalisations 58
Sweden 102, 117

terrorism 118, 125
Thailand 62
training *see education and training*
transport 4, 8, 10, 86, 189, 196

United Kingdom 14, 19, 27, 37, 115,
    118-20, 123, 126, 169-81
United Nations 118
United States of America 75-6, 81,
    117-19, 122, 126, 200

Wageningen 190
Wales 15, 37, 77, 117, 122, 169-81
walking *see hiking*
welfare 76, 90-92, 95, 97-101
wilderness 10, 56, 93
wildlife *see animals*
World Heritage Sites 196
World Tourism Organisation 15, 53